Disclaimer

The publisher of this book is by no way associated with the National Institute of Standards and Technology (NIST). The NIST did not publish this book. It was published by 50 page publications under the public domain license.

50 Page Publications.

Book Title: Agent Stability Under Storage (NIST SP 890)

Book Author: Richard H. Harris;

Book Abstract: Significant losses in fire suppression effectiveness and increases in toxicity are possible if a fire extinguishing agent degrades during multi-year storage. Halon 1301 is known to be stable in metal containers for many years, and any trace degradation products do not affect its fire suppression effectiveness. For candidate replacement agents, comparable data are needed, reflecting the storage conditions of elevated temperature and pressure. The storage environment fosters conditions which may have an adverse effect on the stability of halon replacements. Stored chemicals may engage in oxidation-reduction reactions, hydrolysis, and other corrosive interactions with metal cylinders. They are also subject to unimolecular decomposition and attack by reactive impurities in the agent. Water and oxygen, for example, will sorb to surfaces of cylinders and transfer lines and can never be completely excluded. These sources of instability, along with the possibility of catalytic interactions with the cylinder walls, can promote the evolution of undersirable products and a concomitant loss of fire suppression effectiveness. Toxicity and corrosiveness are particularly important concerns with respect to halogenated compounds, due to the tendency to librate hydrogen halide in the process of degradation. This report gives the details of the test procedure and a comparison of agent absorbance band areas from low concentration spectra. Also, presented are a comparison of absorbance bands in high density spectra for impurities present in the agents or produced as a result of degradation. These data provide a quantification of any degradation of the agents during long-term storage.

Citation: NIST SP - 890

Keyword: fire suppression; aircraft engines; nacelle fires; simulation; storage stability; FT-IR; copper; degradation; halon 1301; halon alternatives; unimolecular decomposition

7. AGENT STABILITY UNDER STORAGE

Richard H. Harris, Jr.
Building and Fire Research Laboratory

Contents

	Page
7. AGENT STABILITY UNDER STORAGE	249
7.1 Introduction	250
7.2 Experimental	251
7.2.1 Test Matrix	251
7.2.2 Materials	251
7.2.3 Preparation of Cylinders for Testing	252
7.2.4 Spectral Analysis Equipment and Procedure	256
7.2.5 Use of the FTIR Spectra to Determine Agent Degradation	256
7.2.6 Determination of Uncertainty	266
7.3 Results	269
7.3.1 FC-218	269
7.3.2 HFC-125	269
7.3.3 HFC-227ea	281
7.3.4 Summary of Fluorocarbon Area Data	281
7.3.5 Iodotrifluoromethane (CF_3I)	282
7.3.5.1 Changes in the Agent Peak Area of CF_3I in the 53.3 Pa Spectra for the Different Conditions	290
7.3.5.2 Changes in the Impurity Peak Areas Appearing in the 5330 Pa Spectra	304
7.3.6 Summary of the CF_3I Area Data	305
7.4 Conclusions	338
7.5 Acknowledgments	338
7.6 References	338
Appendix A. Initial and Final FTIR Spectra for HFC-125	340
Appendix B. Initial and Final FTIR Spectra of HFC-227ea	349
Appendix C. Initial and Final FTIR Spectra of CF_3I Tested in the Dry Condition without Copper	362
Appendix D. Initial and Final FTIR Spectra of CF_3I Tested in the Dry Condition with Copper	378
Appendix E. Initial and Final FTIR Spectra of CF_3I Tested in the Moist Condition without Copper	390
Appendix F. Initial and final FTIR Spectra of CF_3I Tested in the Moist Condition with Copper	397

7.1 Introduction

Significant losses in fire suppression effectiveness and increases in toxicity are possible if a fire extinguishing agent degrades during multi-year storage. Halon 1301 is known to be stable in metal containers for many years, and any trace degradation products do not affect its fire suppression effectiveness. For candidate replacement agents, comparable data are needed, reflecting the storage conditions of elevated temperature and pressure.

The storage environment fosters conditions which may have an adverse effect on the stability of halon replacements. Stored chemicals may engage in oxidation-reduction reactions, hydrolysis, and other corrosive interactions with metal cylinders. They are also subject to unimolecular decomposition and attack by reactive impurities in the agent. Water and oxygen, for example, will sorb to surfaces of cylinders and transfer lines and can never be completely excluded. These sources of instability, along with the possibility of catalytic interactions with the cylinder walls, can promote the evolution of undesirable products and a concomitant loss of fire suppression effectiveness. Toxicity and corrosiveness are particularly important concerns with respect to halogenated compounds, due to the tendency to liberate hydrogen halide in the process of degradation.

Section 6 of NIST Special Publication 861 (Peacock *et al.*, 1994) detailed a screening test for the stability of thirteen agents and eight metals, with the purity of the agent determined by infrared spectral analysis. This test involved storing the agents in PTFE-lined cylinders for one month at a temperature of 150 °C (300 °F). The infrared spectra of the original and "aged" samples were compared. The results of the aging study in combination with other studies of the thirteen agents resulted in the selection of four agents for continued study. These agents were HFC-125, FC-218, HFC-227ea, and CF_3I. Also, in Section 7 of the above publication (Ricker *et al.*, 1994) three of the eight metals studied were chosen as being the best with respect to corrosion. Additionally, the sponsors requested a titanium alloy as a candidate storage metal and copper as an additive for CF_3I be investigated in this study.

Although the screening test was an appropriate method for short-term evaluation, the promising candidate agents need a more rigorous examination. A more comprehensive, long-term evaluation is needed to give stability and degradation information as a function of time and temperature. The information obtained in this study, in combination with information obtained in the studies described in Sections 5 and 6, can then be used in selecting the most thermally stable agents and/or appropriate storage materials.

In this study, samples of the four candidate agents with each of the four storage metal candidates and the copper additive in CF_3I were evaluated in pressurized cylinders at various temperatures and conditions. In order to allow for potential interactions analogous to actual storage conditions, a measured amount of metal (with separate tests for each candidate cylinder metal) was introduced into the containers prior to the experiments. The vessel and its contents were stored at ambient conditions or in ovens at elevated temperatures for as many as 52 weeks as the project time allowed. After specified aging times the cylinders were removed from their respective environments, cooled to ambient conditions, and an infrared (IR) spectrum of the aged sample was collected. Degradation of the sample would be indicated by a systematic decrease in the absorbance of peaks attributable to the agent and/or the appearance of new peaks in the IR spectrum of the aged agent.

This report gives the details of the test procedure and a comparison of agent absorbance band areas from low concentration spectra. Also, presented are a comparison of absorbance bands in high density spectra for impurities present in the agents or produced as a result of degradation. These data provide a quantification of any degradation of the agents during long-term storage.

7. AGENT STABILITY UNDER STORAGE

7.2 Experimental

7.2.1 Test Matrix. Table 1 shows the test matrix. This matrix was designed to utilize oven space, increase the number of exposure temperatures, and expand the exposure conditions to include moisture and addition of copper to CF_3I. The sponsors designed this test matrix to include all agents, all metals, and the various exposure temperatures. The resulting test matrix required 68 sample cylinders. This test procedure was agreed to by NIST.

Since the agents might interact with the metal storage cylinder during long term storage, it was important to evaluate the selected agents in the best metals found in the previous screening study. Containers constructed of each metal would have been ideal for this study, but the cost and availability of cylinders necessitated using metal coupons inside an inert cylinder. In this approach, an amount of metal with the surface area roughly equal to the storage cylinder was introduced into PTFE-lined stainless steel cylinders.

The manufacturers of organoiodide compounds have routinely stored these compounds in the presence of copper (Dierdorf, 1995.) Iodine, if formed during storage, will react with copper to produce cuprous iodide, which is virtually insoluble in most solvents. The sponsors requested that copper be tested as an inhibitor in the CF_3I agent. With this in mind, the test matrix incorporated the addition of copper coupons in certain cylinders, in addition to the test metals.

Since CF_3I appeared to be the least stable agent in the previous screening study, the addition of small amounts of water to some of the test cylinders was incorporated into the test matrix for this agent. In any realistic filling method of storage containers, the presence of residual water in lines and other components of a filling apparatus can be expected. The sponsors requested that the study include the addition of microliter quantities of water into certain cylinders to determine what affect, if any, water has on the degradation of CF_3I.

The temperature dependence of the degradation was also incorporated into the test matrix. This temperature dependence can be dramatic, especially for catalytic degradation. If by-products do form then knowledge of the amounts produced from degradation at different temperatures can be used extrapolate to longer storage times.

7.2.2 Materials. The four agents that were tested for their long-term storage stability are listed in Table 2. They were typical of production grade rather than ultra-pure research samples. All agents were used as received from the manufacturers. Different lots of FC-218 (R-218[1]), HFC-227ea (FM-200), and CF_3I (Triodide) were used in the study. The lots of each agent were compared to each other by comparing their high density Fourier transform infrared (FTIR) spectra (5330 Pa). No differences were found in the lots of FC-218 or HFC-227ea. However, CO_2 was found in the lot of HFC-125. CO_2 and CF_3H were found in the lots of CF_3I. More detail on the amounts of impurities in the lots will be given in 7.2.5.

The five metals that were exposed to each of the agents were the same as those used in the study described in Section 5, Table 1. They were Nitronic 40 (N40), Ti-15-3-3-3 (Ti), C4130 alloy steel (C4130), Inconel 625 (I625), and CDA 110. The metals were supplied by Metal Samples Co., Inc. as coupons 10.2 cm long, 0.8 cm wide, and 0.2 cm thick. In this study, the C4130 alloy steel coupons had a small hole drilled into one end. They were then sent to Walter Kidde Aerospace for phosphate

---

[1]Certain trade names and company products are mentioned in the text or identified in an illustration in order to specify adequately the experimental procedure and equipment used. In no case does such identification imply recommendation or endorsement by the National Institute of Standards and Technology, nor does it imply that the products are necessarily the best available for the purpose.

Table 1. Test matrix

Agent/Test Condition	Metal/Temperature				
	Blank	Nitronic-40	Ti-15-3-3-3	C4130	Inconel 625
FC-218	1T	1T	1T	1T	1T
HFC-125	2T	2T	2T	2T	2T
HFC-227ea	3T	3T	3T	3T	3T
CF_3I/a_1	3T*	3T*	3T*	3T*	3T*
CF_3I/b_1	3T*	2T	2T	2T	2T
CF_3I/a_2	2T*	1T	1T	1T	1T
CF_3I/b_2	2T*	1T	1T	1T	1T

Key:

a = w/o Cu	T_1 = 23 °C	1T = T_4
b = w/Cu	T_2 = 100 °C	2T = T_1, T_4
$_1$ = dry	T_3 = 125 °C	2T* = T_2, T_4
$_2$ = moist	T_4 = 150 °C	3T = T_1, T_3, T_4
		3T* = T_1, T_2, T_4

Table 2. Agents used in agent stability study

Agent	Supplier	Lot Numbers
R-218 (FC-218)	3M Chemolite Center 10	L-12677, L-13201
HFC-125	Allied Signal Chemicals	835
FM-200 (HFC-227ea)	Great Lakes Chemical Corp.	92-002-356, 93-200-278
Triodide (CF_3I)	Deepwater	224940901, 226941712, 226941891

treatment prior to being introduced into the cylinders. The phosphate treatment renders the C4130 alloy steel less susceptible to rusting.

The storage cylinders were constructed of stainless steel, lined with polytetrafluoroethylene (PTFE), and had a 1000 ml capacity. All new cylinders were assembled with high temperature/high pressure stainless steel valves and end plugs, then heated to 150 °C (300 °F) for at least 48 hours with the valves completely open. For re-used cylinders, the heating time was cut to 24 hours.

7.2.3 Preparation of Cylinders for Testing. Each agent/metal combination was exposed to the various temperatures specified in the test matrix. Testing conditions dictated that the maximum exposure temperature be no higher than 150 °C, and that the room temperature fill pressure be

7. AGENT STABILITY UNDER STORAGE

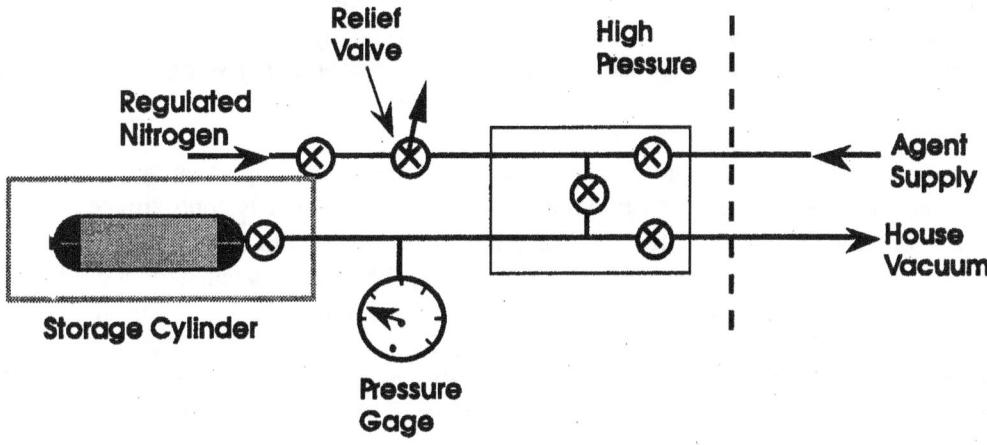

Figure 1. Experimental apparatus for agent filling study.

4.13 MPa (600 psia). In order to assure that the storage cylinder pressure not exceed the pressure rating on the cylinder or valve when exposed to the elevated temperature, a conservatively low amount of agent was placed in each cylinder. The cylinders were filled with agent up to a maximum pressure of the saturation vapor pressure of the agent at room temperature. Nitrogen was then added to achieve the final pressure. Since small amounts of additional liquid would vaporize at the elevated temperature only a single phase would exist anyway, though a significant pressure increase would be observed. The effect of the system pressure on degradation was not explored, but is thought to be small.

To assure that the metal coupons were free of oil and markings, they were immersed and stirred in dichloromethane solvent. The coupons were removed from the solvent, stripped with clean paper towels, and dried at room temperature. The phosphate-treated C4130 alloy steel coupons were used as received from Walter Kidde Aerospace. After cleaning, care was taken to handle the coupons with gloved hands only.

A filling procedure was designed to insure low levels of contaminants (including room air) and a reasonably accurate measurement of the amounts of material introduced into the cylinder. Highly accurate measurement of amounts of agent and/or nitrogen were not required in this method, since it is the change in spectra that was desired. Figure 1 is a schematic of the filling apparatus. It consisted of a three-valve manifold, an absolute pressure gage, and filling lines, all constructed from stainless steel and capable of handling high pressures. The lines to the agent supply tank and the house vacuum were not high pressure rated and were isolated by valves. The following steps outline the filling procedure.

1. Thirty pieces of each metal were weighed and placed into their respective cylinders. The number of coupons was fixed at 30 for each cylinder because that amount is approximately the surface area of the interior of the cylinder. Overlapping of the coupons in the cylinder effectively decreased the exposed surface area, but this decrease, though not calculated, should be similar for all tests, except those containing the additional copper coupons. The end plug threads were wrapped with PTFE tape and tightened. The cylinder valves were opened and the cylinders were heated for at least 12 hours at 150 °C. Immediately, the cylinder valves were closed. When the cylinders were cool enough to be handled with gloves, they were attached to the sampling system used for the FTIR analyses shown in Figure 2. The cylinders were purged with N_2, then evacuated to an absolute pressure of less than 1.3 Pa (0.01 torr). Each evacuated cylinder was weighed.

2. The evacuated storage cylinder was attached to the filling apparatus.

3. Certain cylinders required the addition of water (Table 1). A fitting was designed with a rubber septum. The fitting was attached to the cylinder with N_2 purging the connection. After the fitting was tightened, the cylinder valve was opened. A syringe containing 100 µL of distilled H_2O was inserted through the septum and injected. The cylinder valve was closed and the "evacuated" cylinder was then attached to the filling apparatus.

4. The filling apparatus was evacuated with the house vacuum which was capable of reduced pressures near 34 Kpa. Then the valve to the house vacuum was closed.

5. The filling apparatus was pressurized with nitrogen in excess of 1.4 Mpa. The pressure was released through the relief valve.

7. AGENT STABILITY UNDER STORAGE

6. The filling apparatus was then evacuated again with the house vacuum. The residual room air in the apparatus was less than 1 % of the original amount.

7. The agent inlet valve was opened slowly until the pressure in the filling apparatus reached the vapor pressure of the agent at ambient temperature (the agent inlet line was pre-charged with agent). The agent inlet valve was closed. This dilutes any residual air again. The house vacuum was opened to remove the agent and any remaining residual air. The agent inlet valve was opened slowly until the pressure again reached the agent's vapor pressure. The valve to the evacuated storage cylinder was slowly opened and agent flowed in. Both the inlet valve and the cylinder valve must be opened slowly to avoid the possibility of drawing liquid agent into the apparatus. The system was allowed to equilibrate for one to two minutes, the pressure recorded, and the agent inlet valve closed.

8. The mass of agent in the storage cylinder was determined by removing the first cylinder filled in a series for a given agent and weighing it. The cylinder was then reattached to the filling apparatus. With the cylinder valve closed, the filling apparatus was again evacuated and purged with N_2. Then the cylinder valve was opened and the cylinder was filled to nominally 4.2 Mpa with N_2. This pressure was slightly higher than the prescribed target pressure of 4.13 Mpa. This allowed a small amount of the agent/N_2 mixture to be removed from the cylinder for the initial FTIR analysis. The total pressure was recorded and the valve to the storage cylinder closed.

9. Proceeding with this method for each cylinder was time consuming and introduced the chance of contamination from air. Since the exact amount of agent in each cylinder at the start of aging was not considered critical, each succeeding storage cylinder filled in a series was attached only once and filled with both agent and N_2. The agent pressure was recorded for all cylinders prior to introducing the N_2.

10. The high pressure gas was bled from the filling apparatus and the cylinder disconnected and weighed. At this point, a new cylinder was connected and filled starting with Step 1 above.

11. After FTIR analyses were completed for a series of storage cylinders, they were placed in their proper storage environment as specified in the test matrix. Ambient cylinders were placed on laboratory benches where the temperature remains at 23 ± 1 °C. Cylinders at elevated temperature conditions were placed in laboratory/production ovens with forced convection airflow. The ovens were capable of maintaining a temperature uniformity of ±1 °C at 100 °C and ± 2 °C at 200 °C.

12. When a series of aged cylinders required analysis, they were removed from their respective storage environment. Cylinders at elevated temperature storage were cooled to ambient temperature (at least 2 hours for those at 150 °C). While the cylinders were cooling, the FTIR was prepared for the analyses. After analyses, the cylinders were weighed and the weights recorded. Before being put back into storage, the cylinders were checked for leaks around the valves and endcaps.

13. The cylinders remain in their respective environments pending the desire for longer-term results.

Tables 3-9 list for each cylinder the metal mass, the agent pressure, initial agent/N_2 pressure, and the mass before and after storage for all of the agents as well as for all the different environments of the CF_3I. As mentioned above (steps 8 and 9), only the first cylinder in a given series of agent was actually weighed to see if the agent weight was approximating what was expected from the ideal gas law. For the remaining cylinders in a series the agent mass can be estimated from the agent density at the recorded agent pressure and ambient temperature (along with the fixed 1000 Ml volume of the cylinders. The error in this calculated value is the combination of the uncertainty in the agent pressure reading which is estimated to be ± 0.02 Mpa (± 2.5 psia), deviations of the agent temperature from ambient temperature (at least two minutes were allowed for the system to equilibrate), and the error associated with the equation of state. Here, generalized compressibility charts were used (Balzhiser et al., 1972). Where the tables have a mass value in parentheses the weight was calculated from the measured vapor pressure. The mass before storage was the cylinder weight after the initial spectrum had been taken and the cylinder was ready to go into storage. The mass after storage was the weight of the cylinder after the final spectra were obtained. The difference in the above two masses is thus the amount of agent/N_2 lost as a result of analyses and leakage. Leakage was a problem over the course of the study for roughly 17 of the 68 cylinders in the study. The leakage resulted primarily from cylinders that were reused. The overall validity of the test results was not affected.

7.2.4 Spectral Analysis Equipment and Procedure. The IR analyses were done with a Galaxy Series 7000 FTIR Spectrometer fitted with a KBr beamsplitter and a narrow band mercury cadmium telluride (MCT) detector. The spectral range was from 600 to 4000 cm^{-1} with a resolution of 1 cm^{-1}. A variable path 20-meter gas cell with a volume of 5.4 L and a path length of 10 m was used. A diagram of the sampling system used for the FTIR analyses is shown in Figure 2. All cylinders were analyzed after cooling to ambient temperature. The gas cell was at 105 ± 1 °C for all analyses. The cylinders were connected to the inlet of the gas cell on the FTIR spectrometer. The lines and gas cell were purged with gaseous N_2 from a liquid nitrogen cylinder. A vacuum of less than 1.3 Pa (0.01 torr) was drawn on the gas cell and inlet line up to the cylinder valve. A positive shutoff toggle valve and a micro-metering valve were placed in line between the cylinder valve and the gas cell to accurately control the flow of the agent into the gas cell. The micro-metering valve allowed precise filling of the gas cell and provided reproducible filling of the gas cell. The agent/N_2 was introduced into the gas cell to an absolute pressure of 53.3 ± 0.1 Pa (0.400 ± .001 torr). A spectrum was taken, then the cell was emptied. Two more replicates were done in the same manner. After the third replicate additional agent/N_2 was added to fill the gas cell to 5333 ± 13 Pa (40.0 ± 0.1 torr). A spectrum was taken. The cylinder valve was closed tightly. Then the gas cell and lines were emptied with vacuum and N_2 purging. The cylinders were then removed, the end caps fastened tightly, and weighed. Each cylinder was then checked for leaks before being put back into its aging temperature environment.

The fluorocarbon agents were analyzed every 8 weeks, unless the results indicated no changes were occurring. The CF_3I samples were analyzed every 4 weeks, unless the results indicated no changes were occurring. Samples that were not changing (e.g., those at ambient conditions) were analyzed every 16 or 8 weeks for fluorocarbon and CF_3I cylinders, respectively.

7.2.5 Use of the FTIR Spectra to Determine Agent Degradation. The FTIR spectra of the agents were used as both quantitative and qualitative tools to determine if there was degradation of the agent. As a quantitative tool, an absorbance band for each agent was selected in the 53.3 Pa spectra and the area under this absorbance band was determined for each 4 or 8 week interval. Chemical analysis by IR spectroscopy is based on the assumption that the IR spectrum of a compound is sufficiently unique to identify it. An IR spectrum is obtained by measuring the ratio of the intensity of IR radiation which passes through the sample, I, to the intensity of the incident radiation, I_o, as a

7. AGENT STABILITY UNDER STORAGE

Table 3. Agent and metal amounts in storage cylinders for FC-218

Metal	Temp. (°C)	Metal mass (±0.1 g)	Agent		Agent/N$_2$		
			Pressure (± 0.02 Mpa)	Mass (g)	Pressure (± 0.02 Mpa)	Mass before storage (± 0.1 g)	Mass after storage (± 0.1 g)
Blank	150	--	0.85	79.9	4.20	122.2	95.6
N40	150	344.4	0.84	(82.3)	4.20	120.5	107.5
Ti	150	188.3	0.83	(81.3)	4.24	113.8	76.9
C4130	150	304.5	0.76[a]	(74.5)	4.20	113.0	106.2
I625	150	326.9	0.83	71.9	4.24	118.1	108.7

[a] Storage tank was almost empty when this cylinder was filled.

Table 4. Agent and metal amounts in storage cylinders for HFC-125

Metal	Temp. (°C)	Metal mass (±0.1 g)	Agent		Agent/N$_2$		
			Pressure (± 0.02 Mpa)	Mass (g)	Pressure (± 0.02 Mpa)	Mass before storage (± 0.1 g)	Mass after storage (± 0.1 g)
Blank	23	--	1.28	80.2	4.24	119.7	116.0
N40	23	345.1	1.28	(76.8)	4.24	115.5	109.5
Ti	23	187.5	1.28	(76.8)	4.24	119.2	114.4
C4130	23	304.4	1.28	(76.8)	4.20	121.2	115.2
I625	23	327.0	1.31	78.6	4.27	121.7	114.7
Blank	150	--	1.28	(76.8)	4.27	119.8	111.9
N40	150	343.3	1.28	(76.8)	1.28	117.9	94.6
Ti	150	188.0	1.28	(76.8)	4.27	121.0	105.4
C4130	150	304.2	1.28	78.8	4.24	122.3	114.1
I625	150	327.0	1.28	(76.8)	4.27	118.2	109.0

Table 5. Agent and metal amounts in storage cylinders for HFC-227ea

Metal	Temp. (°C)	Metal mass (±0.1 g)	Agent		Agent/N$_2$		
			Pressure (± 0.02 Mpa)	Mass (g)	Pressure (± 0.02 Mpa)	Mass before storage (± 0.1 g)	Mass after storage (± 0.1 g)
Blank	23	--	0.41	31.8	4.20	77.3	74.6
N40	23	343.7	0.41	(31.8)	4.20	73.7	71.4
Ti	23	187.3	0.41	(31.8)	4.20	76.0	73.2
C4130	23	304.3	0.41	(31.8)	4.20	72.1	68.9
I625	23	327.0	0.41	30.6	4.20	72.0	67.4
Blank	125	--	0.41	32.9	4.20	80.7	74.7
N40	125	344.5	0.41	(31.8)	4.20	74.3	63.1
Ti	125	187.7	0.41	(31.8)	4.20	74.9	64.4
C4130	125	303.9	0.41	(31.8)	4.20	74.6	67.4
I625	125	327.2	0.41	(31.8)	4.20	73.1	68.1
Blank	150	--	0.41	(31.8)	4.20	76.5	71.7
N40	150	344.6	0.41	(31.8)	4.20	72.9	66.1
Ti	150	187.3	0.41	(31.8)	4.20	78.1	70.3
C4130	150	304.4	0.41	(31.8)	4.20	73.9	66.3
I625	150	327.0	0.43	(33.4)	4.20	58.7	42.9

7. AGENT STABILITY UNDER STORAGE

Table 6. Agent and metal amounts in storage cylinders for dry CF_3I samples w/o copper

Metal	Temp. (°C)	Metal mass (±0.01 g)	Agent		Agent/N_2		
			Pressure (± 0.02 Mpa)	Mass (g)	Pressure (± 0.02 Mpa)	Mass before storage (± 0.1 g)	Mass after storage (± 0.1 g)
Blank	23	--	0.45	37.4	4.24	85.4	70.4
N40	23	345.7	0.45	(40.2)	4.24	82.5	71.4
Ti	23	188.3	0.45	(40.2)	4.24	80.1	72.0
C4130	23	304.3	0.45	36.9	4.20	81.4	72.9
I625	23	326.8	0.48	36.8	4.20	79.9	72.4
Blank	100	--	0.48	36.9	4.20	83.3	35.3
N40	100	343.7	0.48	(42.9)	4.20	80.9	--
Ti	100	187.3	0.48	(42.9)	4.20	78.1	70.2
C4130	100	304.3	0.45	(40.2)	4.20	79.1	54.2
I625	100	327.5	0.45	(40.2)	4.20	78.0	71.5
Blank	150	--	0.47	40.1	4.20	87.3	33.8
N40	150	345.1	0.47	(42.0)	4.20	80.9	55.8
Ti	150	187.5	0.47	(42.0)	4.20	84.8	30.1
C4130	150	304.5	0.45	(40.2)	4.20	79.1	32.8
I625	150	327.0	0.48	(42.9)	4.20	80.4	66.3

Table 7. Agent and metal amounts in storage cylinders for dry CF_3I samples with copper

Metal	Temp. (°C)	Cu(CDA110)/ metal mass (±0.1 g)	Agent		Agent/N_2		
			Pressure (± 0.02 Mpa)	Mass (g)	Pressure (± 0.02 Mpa)	Mass before storage (± 0.1 g)	Mass after storage (± 0.1 g)
Blank	23	353.9/--	0.45	(40.2)	4.24	80.4	72.3
N40	23	354.5/342.6	0.45	(40.2)	4.24	77.7	68.1
Ti	23	354.5/187.1	0.45	(40.2)	4.24	78.7	70.6
C4130	23	354.6/304.2	0.45	(40.2)	4.20	74.8	68.6
I625	23	354.1/327.0	0.48	(42.9)	4.20	75.2	25.7
Blank	100	355.1/--	0.48	(42.9)	4.20	80.5	--
Blank	150	352.9/--	0.47	(42.0)	4.20	84.0	71.8
N40	150	352.7/343.5	0.47	(42.0)	4.20	80.2	60.0
Ti	150	352.5/188.2	0.47	(42.0)	4.20	78.5	67.6
C4130	150	354.1/304.2	0.45	(40.2)	4.20	74.8	49.2
I625	150	353.6/327.0	0.48	(42.9)	4.20	77.1	65.3

Table 8. Agent and metal amounts in storage cylinders for moist CF_3I samples w/o copper

Metal	Temp. (°C)	Metal mass (±0.01 g)	Agent		Agent/N_2		
			Pressure (± 0.02 Mpa)	Mass (g)	Pressure (± 0.02 Mpa)	Mass before storage (± 0.1 g)	Mass after storage (± 0.1 g)
Blank	100	--	0.48	(42.9)	4.20	81.8	73.4
Blank	150	--	0.47	39.1	4.20	86.0	60.1
N40	150	345.5	0.47	(42.0)	4.20	82.4	63.9
Ti	150	188.1	0.47	(42.0)	4.20	80.6	60.7
C4130	150	304.1	0.45	(40.2)	4.20	77.1	63.7
I625	150	327.0	0.48	(42.9)	4.20	80.7	70.5

7. AGENT STABILITY UNDER STORAGE

Table 9. Agent and metal amounts in storage cylinders for moist CF_3I samples with copper

Metal	Temp. (°C)	Cu(CDA110)/ metal mass (±0.1 g)	Agent		Agent/N_2		
			Pressure (± 0.02 Mpa)	Mass (g)	Pressure (± 0.02 Mpa)	Mass before storage (± 0.1 g)	Mass after storage (± 0.1 g)
Blank	100	354.2/--	0.48	(42.9)	4.20	79.1	70.7
Blank	150	354.4/--	0.47	(42.0)	4.20	80.0	54.9
N40	150	353.7/344.4	0.47	(42.0)	4.20	76.6	64.9
Ti	150	352.6/187.4	0.47	(42.0)	4.20	75.4	64.5
C4130	150	353.5/304.3	0.45	(40.2)	4.20	73.0	59.1
I625	150	352.7/319.1	0.45	(40.2)	4.20	73.9	61.9

function of frequency (v). Concentrations of individual components are quantified by applications of Beer's law

$$C_i = \left(\frac{A_i(v)}{A_r(v)}\right) C_r, \qquad A(v) = -\log\frac{I(v)}{I_o(v)}. \tag{1}$$

where subscript r indicates known values obtained from a reference sample. IR spectroscopy is extremely versatile in the sense that almost all compounds, with the notable exception of single atoms and homonuclear diatomics, are IR active. In principle, IR analysis makes it possible to monitor the degradation of each candidate and identify the corresponding products from a simple before and after comparison of the spectra. In practice, IR analysis can suffer from overlapping peaks in a spectrum similar to matrix effects with other analytical methods. In addition, quantitative determination of degradation products require either pre-existing spectra for comparison (libraries of spectra for common materials are commercially available) or available reference samples. The criteria for choosing an agent absorbance band for integration were the following: an absorbance band representing the weakest bond in the molecule; a small band with maximum absorbance less than 0.6 so that if a reduction in area occurred as a result of degradation the change would be detected; assumed to obey Beer's law at lower concentrations if the absorbance is less than 0.6 at higher concentrations; and well resolved with respect to the baseline noise.

FC-218 is a symmetrical molecule with 2 C-C and 8 C-F bonds. The well resolved peak at 731 cm^{-1} was chosen for integration. Both HFC-125 and HFC-227ea contain C-C, C-H and C-F bonds. The C-H stretch absorbance band near 3000 cm^{-1} for each of these agents was chosen. All of these chosen agent absorbance bands satisfied the above criteria. The spectra for these agents are shown in Figures 3-5 with the absorbance band that was integrated noted.

For CF_3I, the ideal absorbance band to have integrated would have been that for the C-I stretch. The fundamental vibrational frequency for the C-I stretch from the literature (McGee, 1952) is 539 cm^{-1}. However, the current detector is only sensitive from 4,000 to 600 wavenumbers. As shown in the lower spectrum in Figure 6, a small, but well resolved peak appears at 2255 cm^{-1}. This peak

262   7. AGENT STABILITY UNDER STORAGE

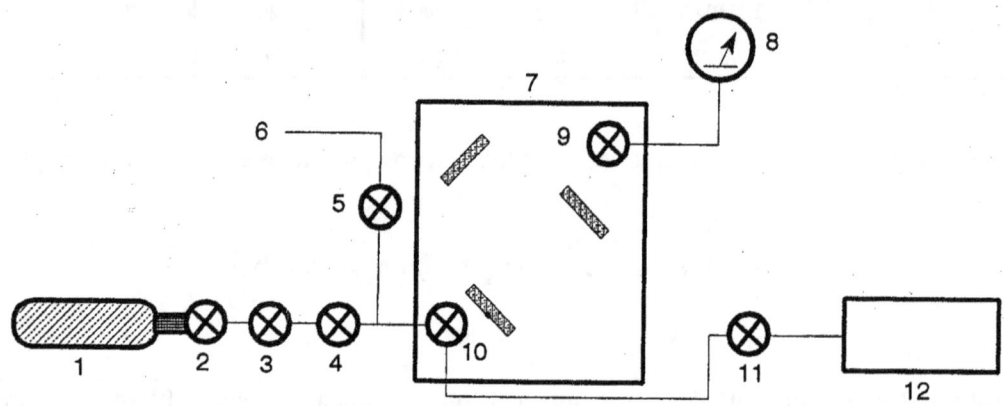

1. Agent Storage Cylinder
2. Cylinder Valve
3. Toggle Valve
4. Micro-Metering Valve
5. Toggle Valve
6. N_2 Purge Line
7. Variable Path Length Gas Cell @ 10 m Path Length
8. Pressure Gauge
9. Gas Cell Pressure Gauge Valve
10. Gas Cell Filling Valve
11. Main Vacuum Valve
12. Two-Stage Vacuum Pump

Figure 2.   Diagram of the equipment for the FTIR spectral analyses.

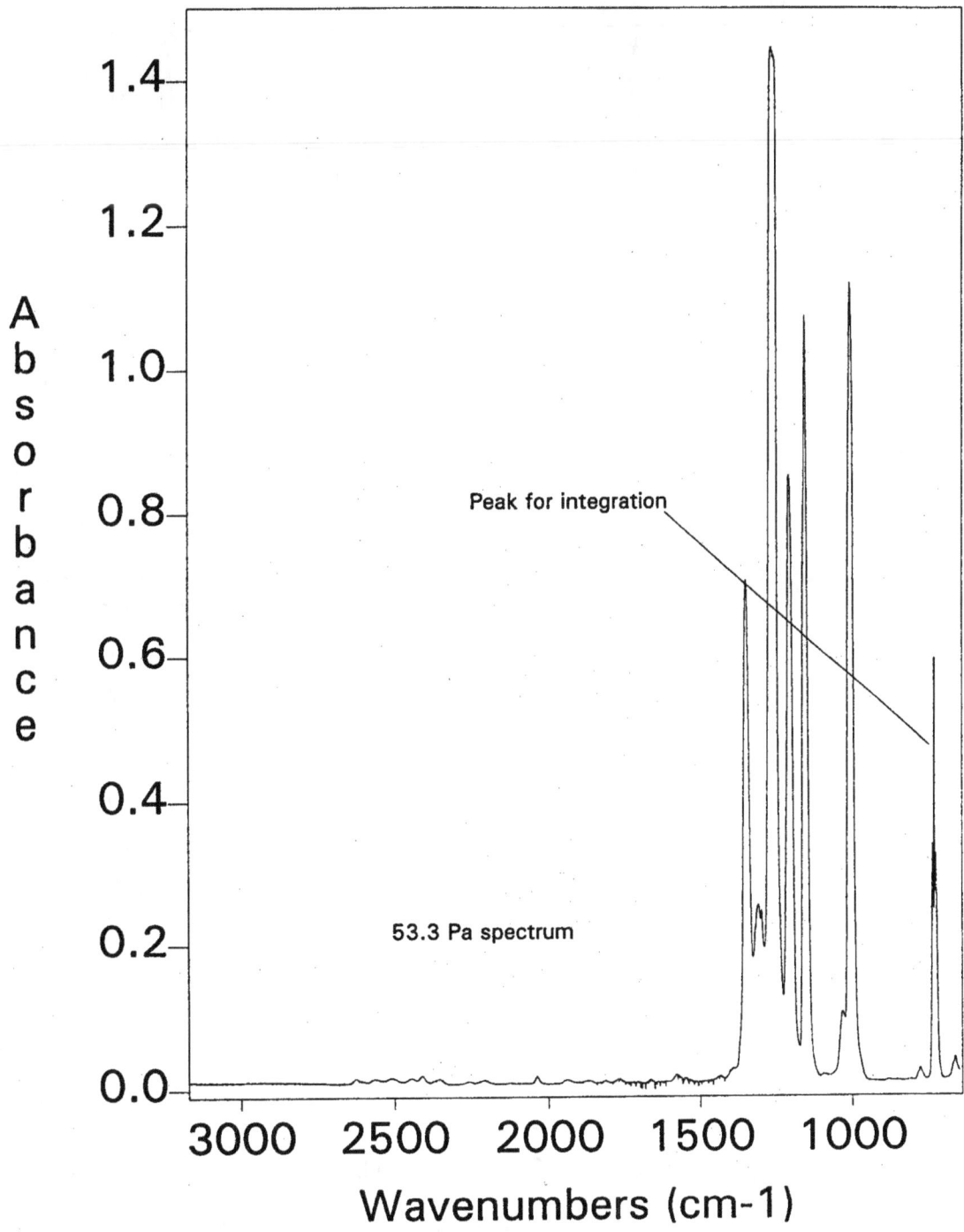

Figure 3. Agent absorbance band in the 53.3 Pa IR spectrum of FC-218 that was integrated.

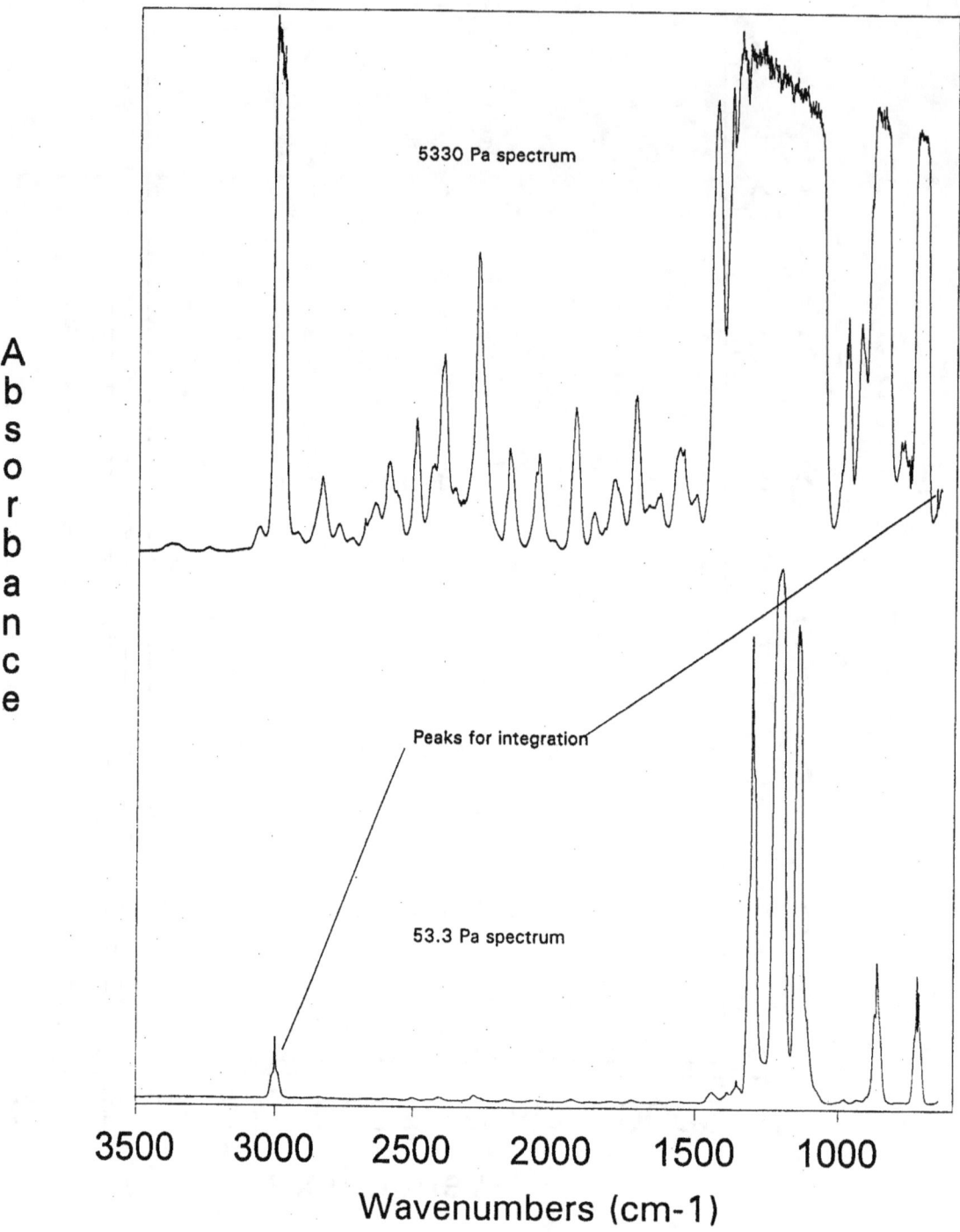

Figure 4. Agent absorbance band in the 53.3 Pa IR spectrum and CO_2 absorbance band in the 5330 Pa IR spectrum of HFC-125 that were integrated.

7. AGENT STABILITY UNDER STORAGE

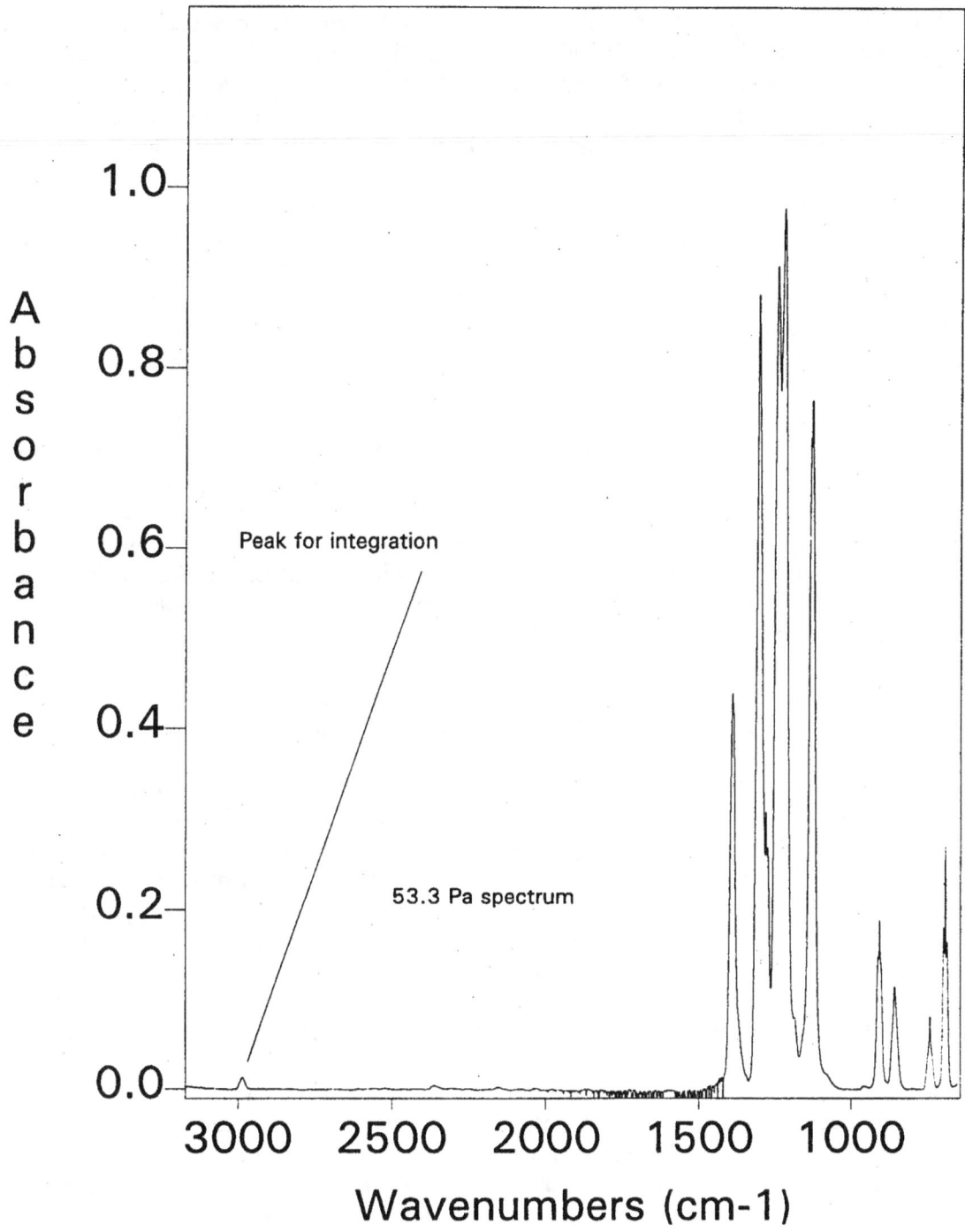

Figure 5. Agent absorbance band in the 53.3 Pa IR spectrum of HFC-227 that was integrated.

corresponds to a combination mode of two larger peaks that correspond to the fundamental stretching frequency of C-F bonds and are labelled peaks 1 and 2 in Figure 6. The small peak thus results from both the stretching and bending of symmetrical C-F bonds. If only a small fraction of CF_3I molecules in the cylinder are degrading, then a reduction in area under this small band will be more easily detected because the low intensity band is more likely to follow Beer's law. Therefore, we used the 2255 cm^{-1} absorbance band.

As a qualitative tool, higher concentration spectra were used. If degradation of the agent were to occur, then the degradation products should begin to form and impurity bands in the aged IR spectrum from each cylinder should appear. Since the amount of degradation products might be small in comparison to the amount of agent, aged spectra at 5330 Pa (or 100 times as dense as the quantitative spectra) were obtained for each cylinder at each testing interval. The baselines of these higher concentration, "qualitative spectra" were examined closely to determine if new absorbance bands were appearing.

In addition to any new bands appearing, bands from impurities already present in the agents could be determined from the higher concentration spectrum. Impurity bands from CO_2 were found in all of the initial spectra at 5330 Pa for both the HFC-125 and CF_3I samples. CO_2 might have been present in FC-218 and HFC-227ea, however, this could not be determined because both agents have strongly absorbing bands in the 670 and 2360 wavenumber regions which obscures the absorption spectra of CO_2. Also detected in the initial spectra for the CF_3I cylinders was CF_3H. Since different lots of some agents were used, Table 10 shows the initial areas for the impurity absorbance bands for each lot of agent determined from the initial 5330 Pa FTIR spectrum.

The only agent to show a new absorbance band appearing in the spectra at elevated temperatures was CF_3I. This new absorbance band will be discussed in 7.3.5.2.

Figures 4 and 6 also show the absorbance bands in the higher concentration spectra that were integrated.

As the study progressed, it became obvious that the 5330 Pa spectra for samples stored at 150 °C were changing, especially for the CF_3I samples. A software program (Mattson Instruments, Inc., 1992) generates a correlation coefficient comparison (Mendenhall *et al.*, 1992) report of sample data (aged spectrum) and reference data (initial spectrum). This spectral comparison takes into account changes that are occurring in a spectrum as a result of changes in existing peaks and the appearance of new absorbance peaks. The more a spectral comparison decreases from 1.000 (the value obtained when a spectrum is compared to itself) represents the magnitude of the changes taking place. Thus the value of the spectral comparison for an aged spectrum compared to its initial spectrum allowed a quantitative value to be assigned to the changing spectra as a function of temperatures and conditions of aging.

7.2.6 Determination of Uncertainty. A Type A evaluation of the standard uncertainty (Taylor, 1995) was used in this study. The samples kept at ambient conditions (23 °C) were not changing with time. This was evident by seeing no pattern in the measured integrated peak areas as a function of time. For each agent and metal combination at 23 °C, the peak area measurements obtained over the study period were averaged and a standard deviation was determined. As a worst-case scenario,

7. AGENT STABILITY UNDER STORAGE

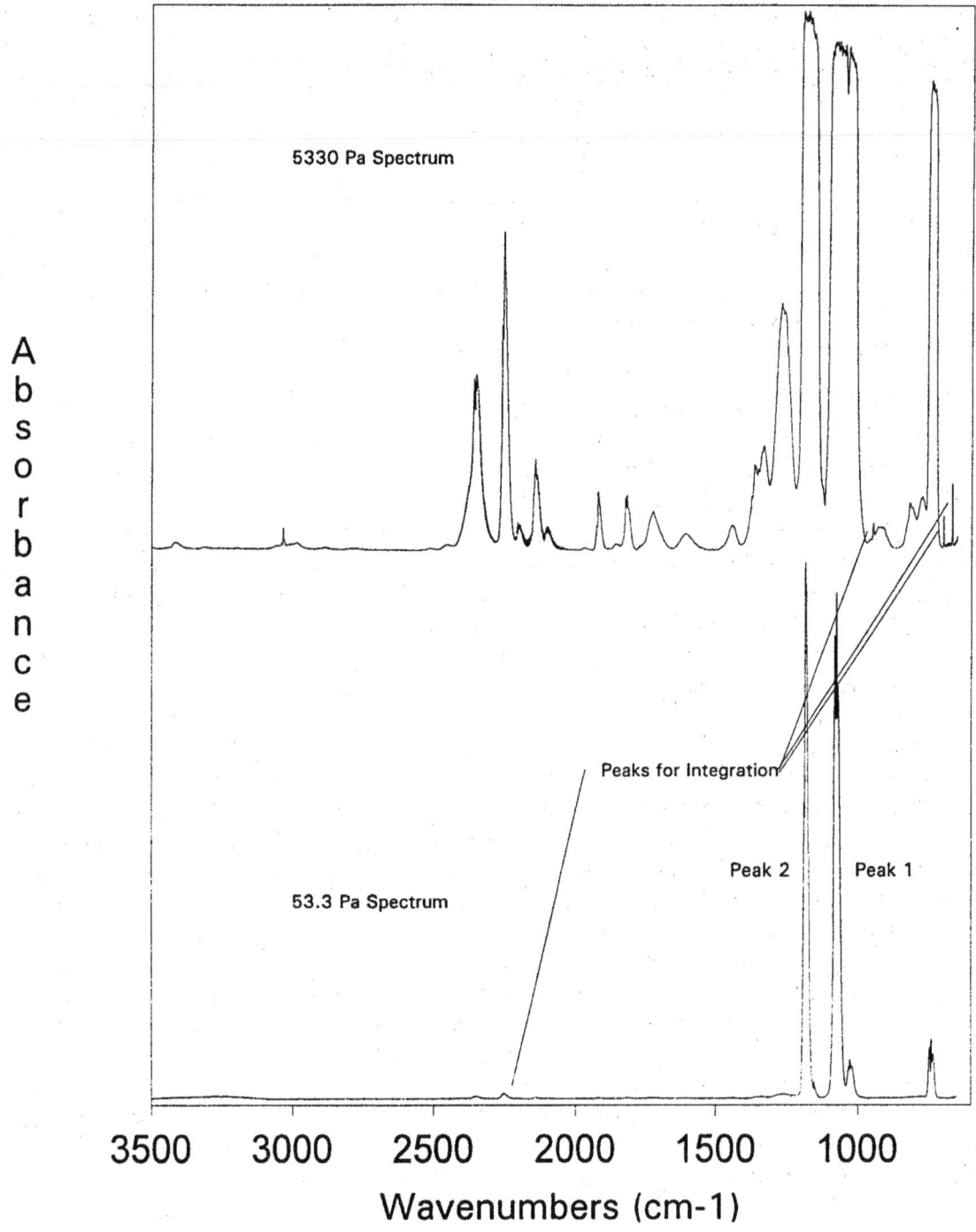

Figure 6. Agent absorbance band in the 53.3 Pa IR spectrum and impurity absorbance bands in the 5330 Pa IR spectrum of CF_3I that were integrated.

Table 10. Impurity levels in the lots of agents at the start of the study

Agent	Lot numbers	Integrated area Impurity compound	
		CO_2	CF_3H
FC-218	L-12677	ND[a]	ND
HFC-125	835	0.01	ND
HFC-227ea	92-002-356	ND	ND
HFC-227ea	93-200-278	ND	ND
CF_3I	224940901[b]	0.09	0.01
CF_3I	226941712[c]	0.18	0.04
CF_3I	226941891[d]	0.01	ND

[a] Not detected
[b] Storage tank was roughly half full when analyzed
[c] Storage tank was full when analyzed
[d] Storage tank was nearly empty when analyzed.

the largest standard deviation for a given agent and condition was used for the uncertainty. Agents FC-218 and CF_3I in the moist condition were exceptions. Neither of these samples had ambient cylinders specified in the test matrix. The uncertainty reported for FC-218 was made from the 150 °C peak area data as stated above, based on the fact that no changes in the 5330 Pa spectra were occurring. The uncertainty for the moist CF_3I samples was assigned the same as that for the dry condition. Table 11 lists the wavenumber ranges over which integration was performed for each agent and impurity band along with the uncertainty in the area measurements and the number of observations (N). Note the relatively high uncertainties for the FC-218 and HFC-125. These were the two strongest absorbance bands in the 53.3 Pa spectra of these agents.

The uncertainty in the spectral comparison values was determined in the same manner as described above for the change in area. For each agent and metal combination at 23 °C, each aged spectrum was compared to its respective initial spectrum. These values were averaged and a standard deviation was determined. Once again the largest standard deviation was chosen as the standard uncertainty. The spectral comparisons and uncertainties are reported below in the tables of agent absorbance band areas for each agent and metal combination at 23 °C, along with the number of observations (N).

In some cases the impurity bands for CO_2 and CF_3H in the initial spectra were very small. To ascertain whether these bands could be measured reliably, a measure of the uncertainty in the detection of these peaks was determined. Examination of the baseline of a typical FTIR spectrum in the expanded axis mode reveals a baseline that is sinusoidal in nature. One of these very small "sinusoidal" peaks where no absorbance was occurring as a result of agent, and near the impurity absorbance band of interest was integrated. One of the smallest areas for absorbance reported (0.01) was three times larger than one of the largest baseline absorbance areas. The changes in the agent absorbance bands and impurity bands are described below for the individual agents.

7. AGENT STABILITY UNDER STORAGE

Table 11.   Wavenumber ranges for integrated areas of agent and extraneous peaks

Agent	Wavenumber range for integrated area (cm^{-1})	Uncertainty in measured area/N
FC-218	710 - 750	0.15/6
HFC-125	2963 - 3038 (Agent) 667 - 671 (CO_2)	0.042/4 0.0023/4
HFC-227ea	2966 - 3006 (Agent)	0.0095/4
CF_3I	2219 - 2274 (Agent) 667 - 671 (CO_2) 698 - 701 (CF_3H) 944 - 953 (F-alkene)	0.011/9 0.0048/9 0.0011/9 0.011/9

Because of the large number of samples, the time required to obtain all materials and equipment, and the time required for initial analyses, the introduction of cylinders into the test matrix took several months. Thus, not all metals in a given agent or test condition of CF_3I received the same time of aging. In the tables of integrated data area, the last column to have data reported was the extent of aging. Because the agents, especially CF_3I, were in short supply at the beginning of this study and other studies had to get underway, it was not possible to retain a particular lot of any agent. When a series of cylinders was filled, the agent storage tank was passed to other researchers. By the time that agent was needed again in this study, all of the lot was sometimes used. Therefore, getting all of the cylinders started required the use of different lot numbers of agents. These factors added to the uncertainty of the data analyses. Operator experience and care in performing the analyses also affected the spectral comparisons. For this reason, an attempt to establish Type B evaluations of standard uncertainty (Taylor, 1995) were not done.

7.3 Results

7.3.1 FC-218. At the onset of this investigation, FC-218 was considered to be the most stable of the four agents. Therefore, the test matrix (Table 1) specified this agent to be tested only at 150 °C. Figures 7-11 show the overlaid spectra of the initial and final aged spectra at 5330 Pa for each metal and blank. Visual examination of these spectra indicated no changes occurred. Table 12 shows the integrated area for the agent absorbance band with time and the spectral comparison for the samples. The initial area value for C4130 is an outlier known to result from improper filling of the IR gas cell. Figure 12 shows the integrated areas in Table 12 graphically represented. The areas tended to drift upward over the 48 week period. This was the only agent for which the integrated area for the absorbance band of an agent showed the slight upward drift. The data for this agent had more scatter than any of the other agents. However, the spectral comparison values in the last column of Table 12 indicate the spectra are unchanged.

7.3.2 HFC-125. Because of the weaker C-H bond in this molecule, the test matrix (Table 1) specified two test temperatures. A typical initial and aged spectra at 5330 Pa and at 23 and 150 °C for the

Table 12. Integrated peak areas for a C-F absorbance band in the 53.3 Pa spectra for FC-218 at 150 °C and the spectral comparisons

Metal	Integrated area from 710 to 750 cm^{-1} (± 0.15) Week number							Spectral comparison (± 0.00070)[a]
	0	8	16	24	32	40	48	
Blank	6.2	6.3	NM[b]	6.4	6.4	6.5	6.5	0.996
N40	6.6	6.8	NM	6.9	6.8	7.0	7.0	0.996
Ti	6.6	6.7	NM	6.8	6.7	6.8	6.7	0.996
C4130	5.7[c]	6.5	6.4	6.5	6.5	6.5	NM	0.999
I625	6.5	6.5	6.7	6.6	6.6	6.7	6.9	0.998

[a] Number of observations was 5
[b] Not measured
[c] Outlier from improper gas cell filling.

blank for HFC-125 are shown in Figures 13 and 14, respectively. Visual examination of the initial and final spectra for the cylinder at 23 °C (Figure 13) showed no change in the absorbance and no new peaks. Visual examination of the initial and final spectra for the cylinder tested at 150 °C (Figure 14) did show a small increase in the absorbance band for CO_2 in the 670 cm^{-1} region. The samples containing each of the four metals gave similar results. Their spectra are shown in Appendix A for reference.

Table 13 lists the integrated areas under the C-H stretch absorbance band with time and temperature for each metal and temperature. Figure 15 is a graphic representation of this data. In all cases the peak areas randomly varied from 2.7 to 2.9. The graphs showed some scatter of the areas, but no upward or downward drift with time. All of the areas reported for the aged samples were within three standard deviations of its initial area. At the 99 % confidence interval, the spectral comparisons did not change significantly either.

7. AGENT STABILITY UNDER STORAGE

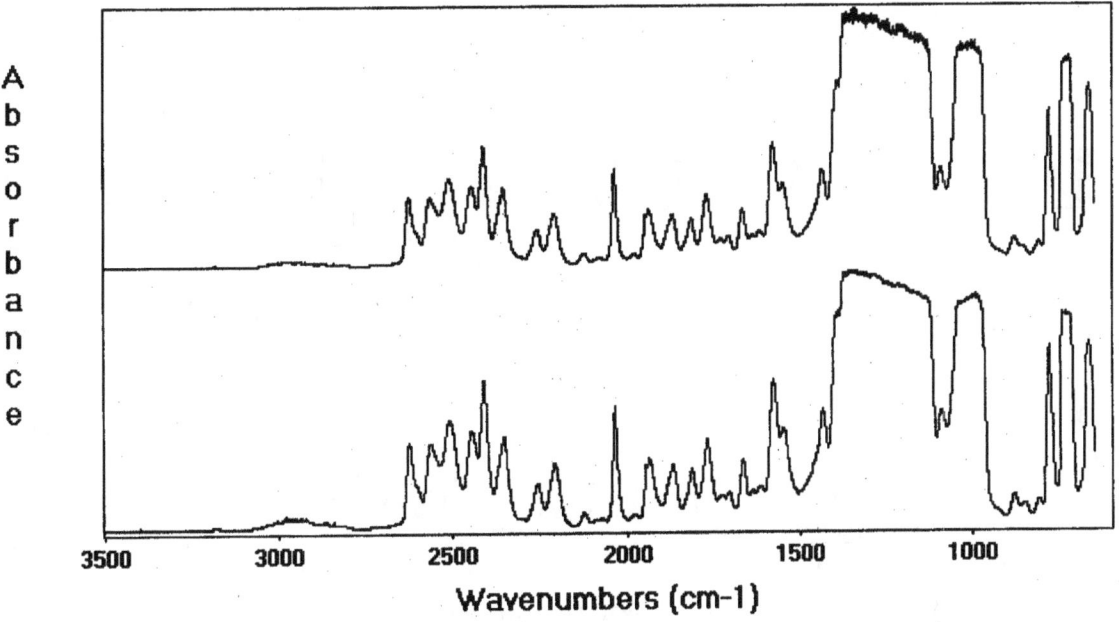

Figure 7. Initial (lower) and 48 week (upper) 5330 Pa spectra for the blank for FC-218 at 150 °C.

272　　　　　　　　　　　　　　　　　　　　　　　7. AGENT STABILITY UNDER STORAGE

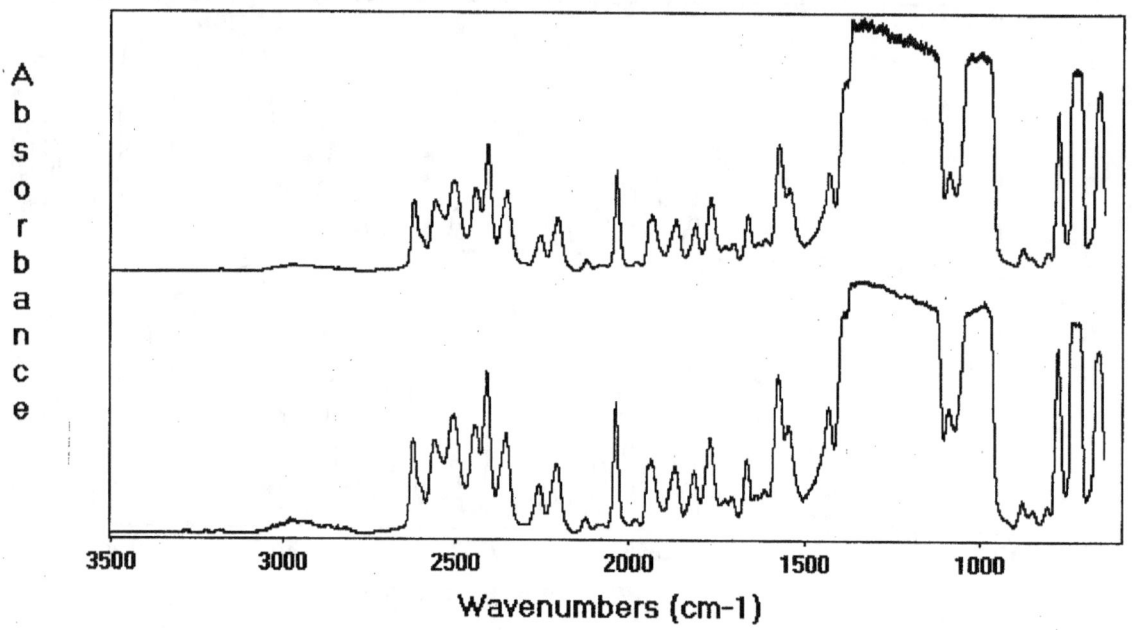

Figure 8.　Initial (lower) and 48 week (upper) 5330 Pa spectra for nitronic 40 in FC-218 at 150 °C.

7. AGENT STABILITY UNDER STORAGE

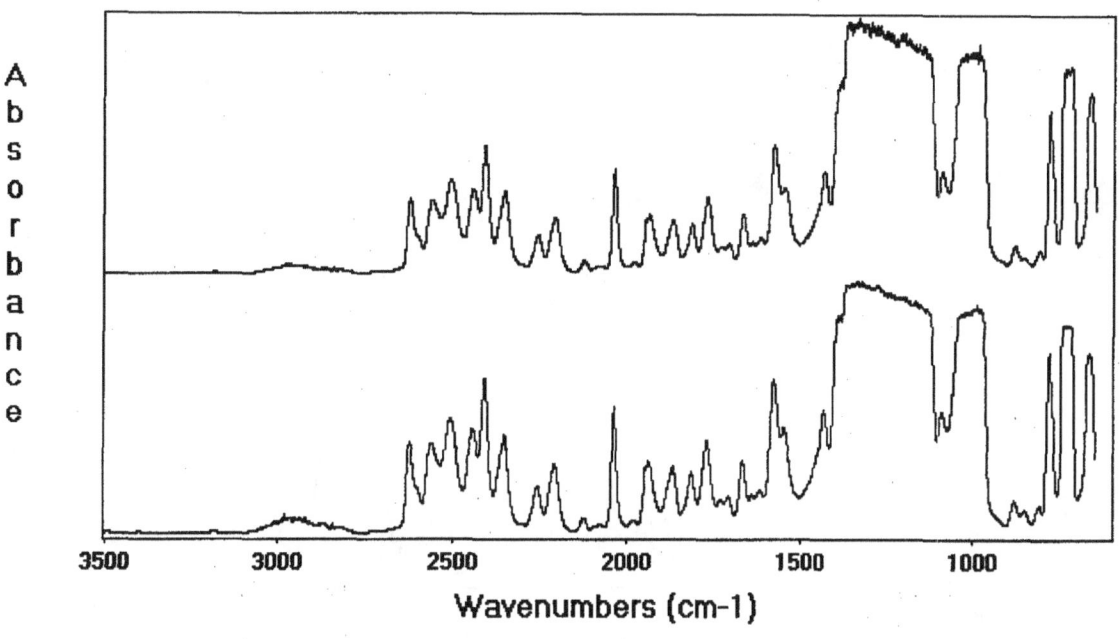

Figure 9. Initial (lower) and 48 week (upper) 5330 Pa spectra for Ti-15-3-3-3 in FC-218 at 150 °C.

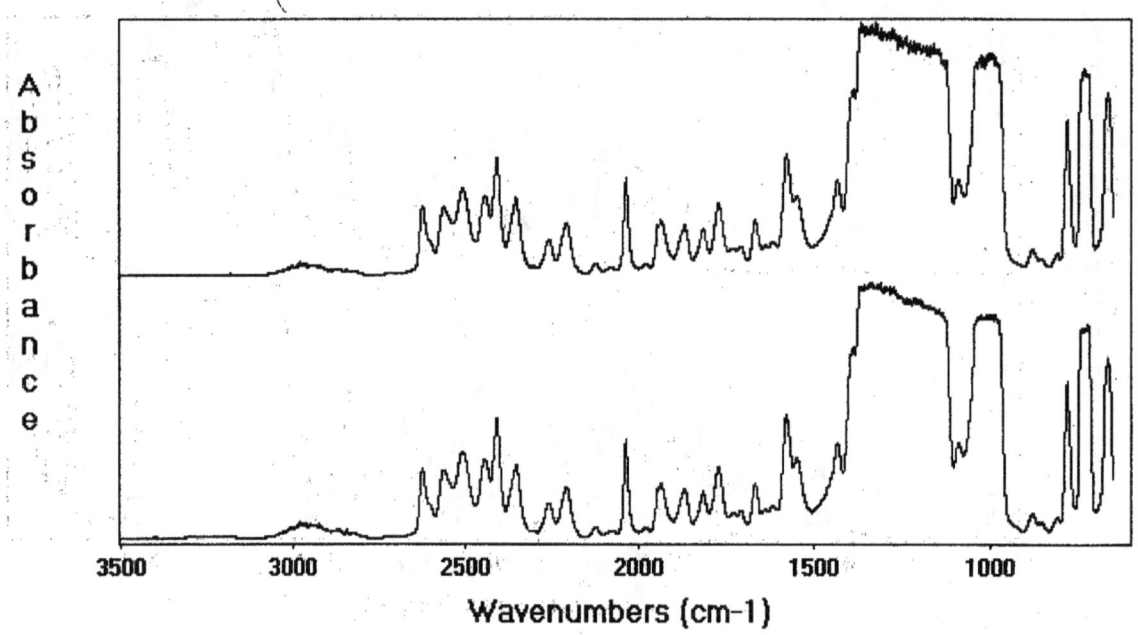

Figure 10. Initial (lower) and 40 week (upper) 5330 Pa spectra for C4130 in FC-218 at 150 °C.

7. AGENT STABILITY UNDER STORAGE

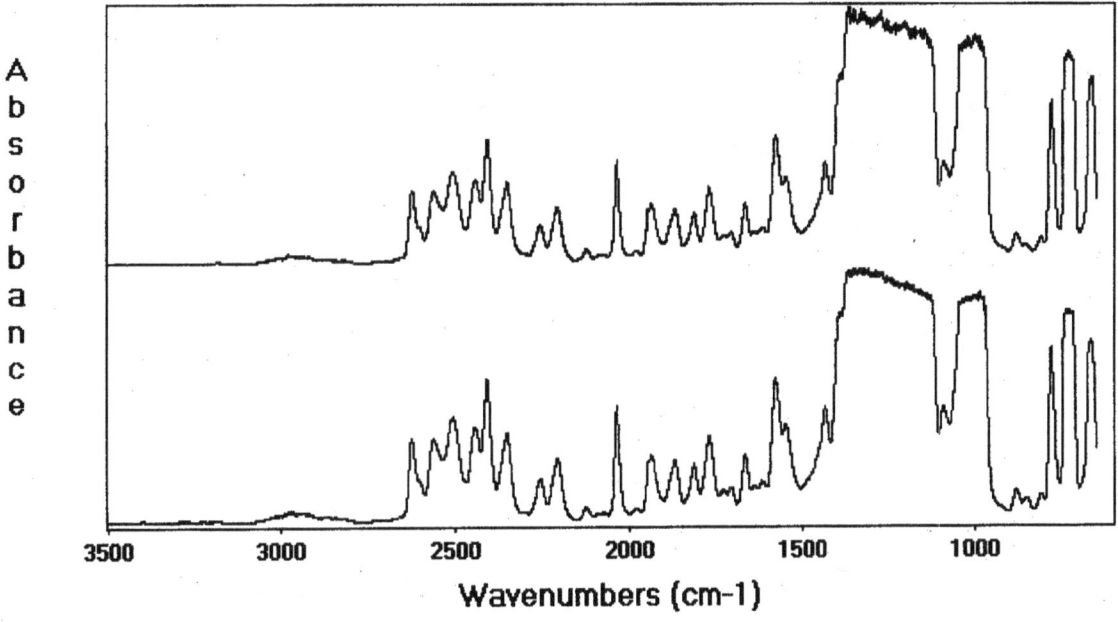

Figure 11. Initial (lower) and 48 week (upper) 5330 Pa spectra for I625 in FC-218 at 150 °C.

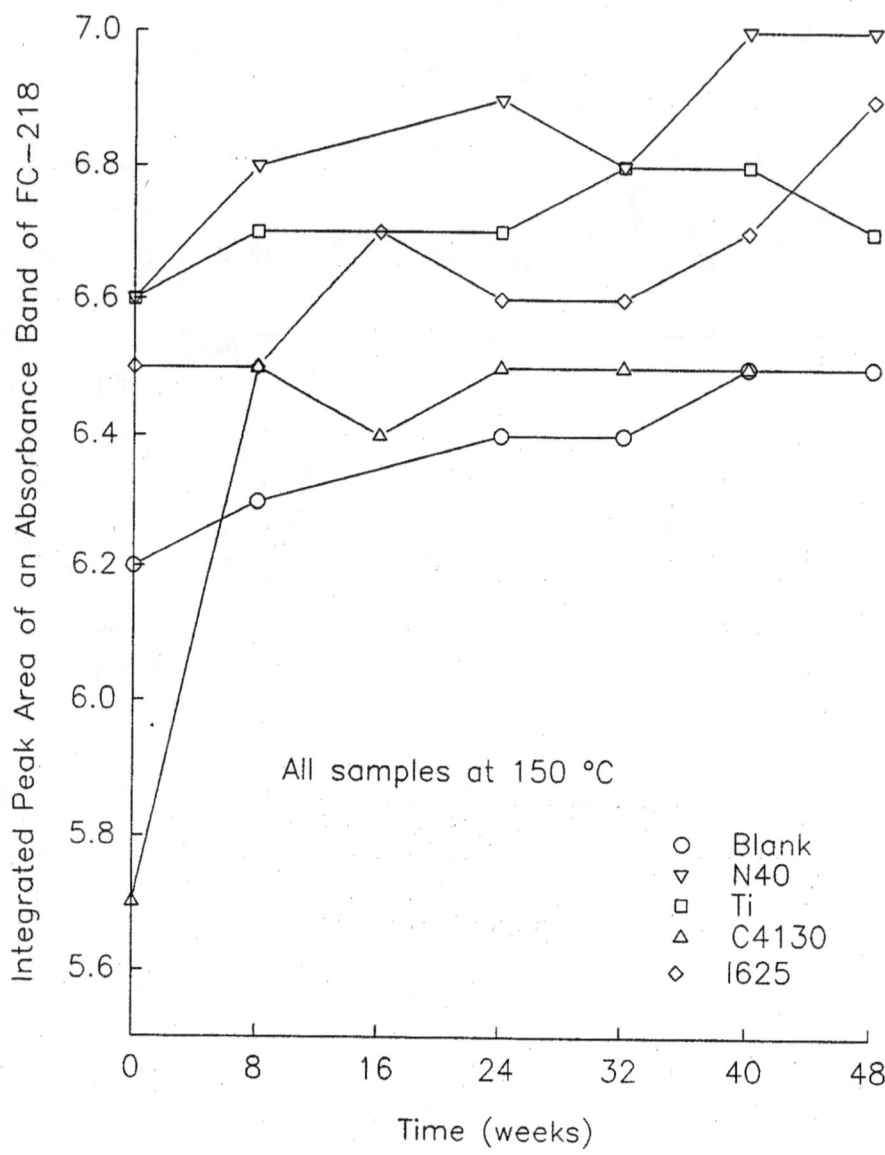

Figure 12. Integrated peak areas for all samples of FC-218 at 150 °C plotted as a function of time.

7. AGENT STABILITY UNDER STORAGE 277

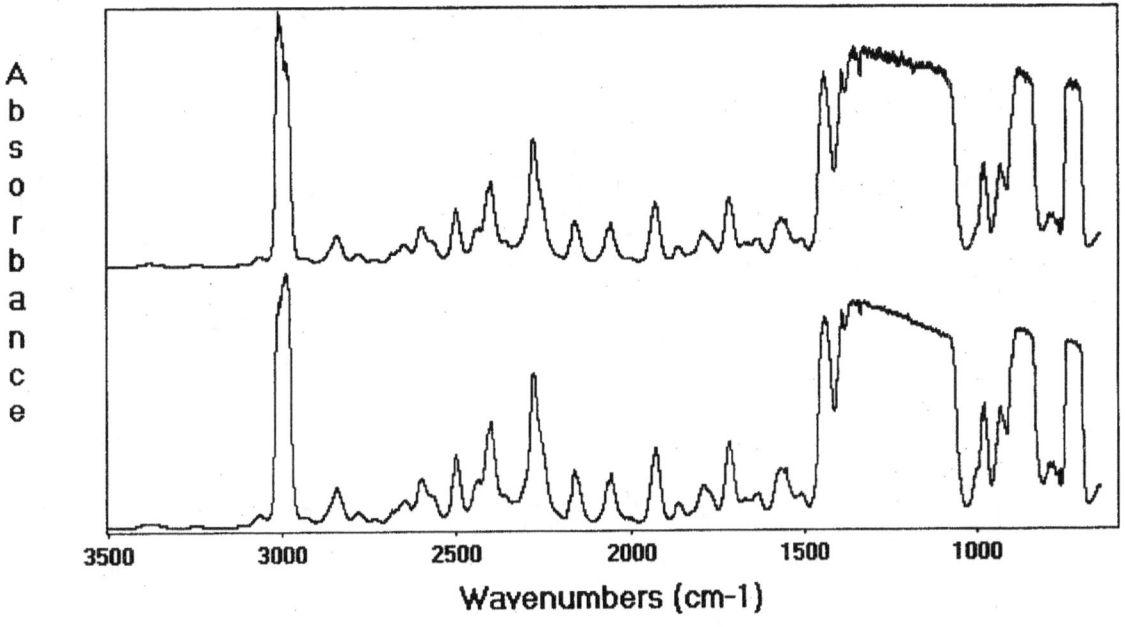

Figure 13. Initial (lower) and 48 week (upper) 5330 Pa spectra for the blank for HFC-125 at 23 °C.

278 7. AGENT STABILITY UNDER STORAGE

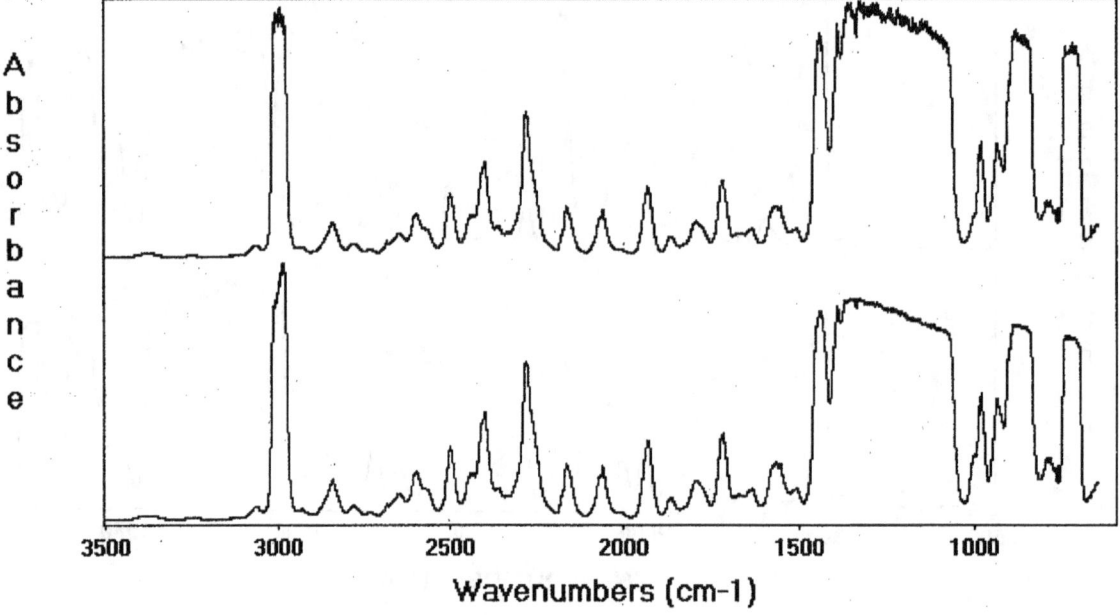

Figure 14. Initial (lower) and 48 week (upper) 5330 Pa spectra for the blank in HFC-125 at 150 °C.

Table 13. Integrated peak areas for the C-H stretch absorbance band in the 53.3 Pa spectra for HFC-125 at 23 and 150 °C and the spectral comparisons

Metal	Temp. (°C)	Integrated Area from 2963 to 3038 cm^{-1} (± 0.042)							Spectral Comparison (± 0.00095)[a]
		Week Number							
		0	8	16	24	32	40	48	
Blank	23	2.8	NM[b]	NM	2.7	NM	2.8	2.7	0.996
N40	23	2.7	NM	NM	2.7	NM	2.7	2.7	0.997
Ti	23	2.8	NM	NM	2.8	NM	2.8	2.8	0.996
C4130	23	2.8	2.8	2.8	2.9	2.8	NM	NM	0.998
I625	23	2.8	NM	2.8	2.8	2.8	2.8	2.9	0.998
Blank	150	2.8	2.8	NM	2.8	2.7	2.8	2.8	0.995
N40	150	2.8	2.8	NM	2.8	2.7	2.8	2.8	0.995
Ti	150	2.8	2.8	NM	2.8	2.7	2.8	2.8	0.996
C4130	150	2.8	2.9	2.8	2.8	2.9	2.9	NM	0.998
I625	150	2.8	2.9	2.8	2.8	2.8	2.8	2.9	0.998

[a] Number of observations was 3
[b] Not measured

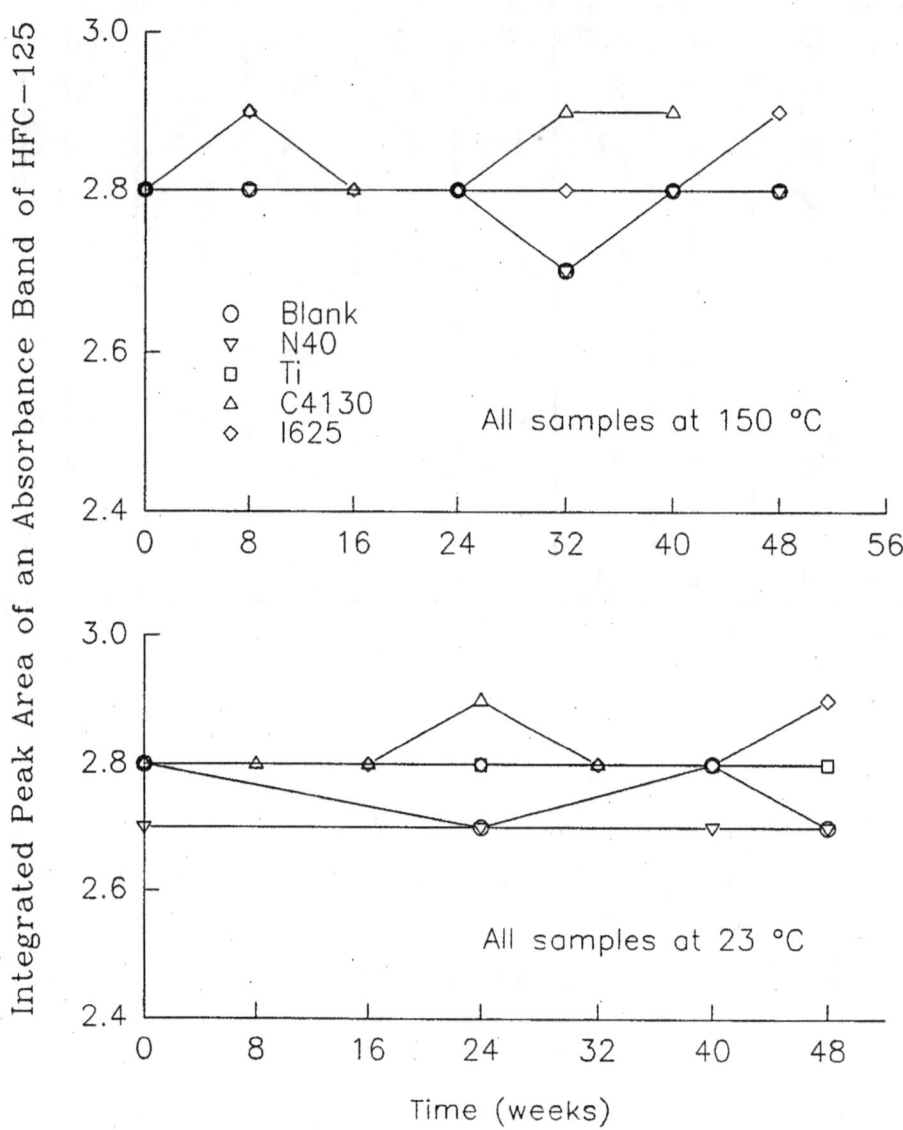

Figure 15. Integrated peak areas for all samples of HFC-125 at 23 °C and 150 °C plotted as a function of time.

7. AGENT STABILITY UNDER STORAGE

In the 5330 Pa spectrum of HFC-125 a small amount of CO_2 (<100 µL/L) was detected in all of the initial analyses. Table 14 shows the integrated areas under the absorbance band at 670 cm^{-1} for CO_2 for all samples and temperatures. Figure 16 is the graphic representation of this data. At 23 °C no changes in the amount of CO_2 were seen. At 150 °C the CO_2 content in the samples increased anywhere from three and a half to a twelve times after 32 weeks. Beyond 32 weeks the CO_2 concentration remained constant. These data confirm the visual observations of an increase in the CO_2 content with time and temperature. The CO_2 could be from outgassing or from agent decomposition in air, which would also yield H_2O, HF, HI, or other compounds. These would be measurable if they didn't sorb to or react with the cylinder walls or metal coupons. There were no hints from the spectra of the above compounds forming, so the CO_2 increase was probably from outgassing. It should be noted that the increase in this peak is not sufficient to change the spectral comparisons.

There are no other apparent changes in the initial and final spectra of any of the cylinders at 23 or 150 °C.

7.3.3 HFC-227ea.
HFC-227ea is a three carbon chain molecule with C-C and C-F bonds and a C-H bond. Of the three fluorocarbon agents this agent was thought to be the most likely to undergo degradation at elevated temperatures. The test matrix (Table 1) specifies that this agent was to be tested at 23, 125, and 150 °C to see if thermal degradation would result. Figures 17-19 show the initial and final spectra at 5330 Pa for the blank at the three temperatures, respectively. Visual examination of the baselines and comparison of the spectra at 5330 Pa show no additional absorbance bands appearing in any of the spectra at any temperature. This is true for all of the metal samples at all three temperatures. The initial and final spectra for all of these samples are shown in Appendix B for reference.

Table 15 shows the integrated peak areas for the C-H absorbance band and the spectral comparisons for each metal and blank at each temperature. The integrated area for I625 at 150 °C is half that for the other cylinders. This is a result of incomplete filling at the outset. Figure 20 is the graphical representation of the data in Table 15. As with the HFC-125, there was little if any change in the peak areas with time or temperature. Also the data show no tendencies to drift either up or down. All of the reported aged data areas fall within two standard deviations of their respective means. The spectral comparisons are all 0.997 or greater. The blank, Nitronic 40, and Ti at 125 °C all had spectral comparisons higher than their respective 23 °C controls. The 125 °C samples were prepared and analyzed later in the study. By this time the analytical technique had become more efficient and quite routine, leading to better spectral comparisons.

7.3.4 Summary of Fluorocarbon Area Data.
The above graphs of the agent peak areas suggested that no changes were occurring in the agents. Table 16 is a summary of the area data from Tables 12, 13 and 15. This table shows only the initial and final peak areas along with the number of weeks each sample was aged. The last column shows whether the final data areas changed significantly at the 99 % confidence interval. None of the final peak areas were significantly less for any agent. All of the final areas for the FC-218 were actually higher than their original peak areas. For a peak to increase, the change has to be in the analysis. As such, the "increase" may set a level of uncertainty.

Table 17 summarizes the spectral comparisons for all agents tested at 150 °C. Since there were no ambient controls for FC-218, the spectral comparisons listed for this agent are simply the final spectra compared to the initial spectra. These spectral comparisons suggest that little change in any of the samples occurred. For the other two agents, the spectral comparisons are amazingly similar. With the exception of the blank and Nitronic 40 in HFC-125, which showed a slight decrease, but not statistically less at the 99 % confidence interval, there were no changes in the spectral comparisons for

Table 14. Integrated peak areas for an absorbance band of CO_2 in the 5330 Pa spectra of HFC-125 at 23 and 150 °C

Metal	Temp. (°C)	Integrated Area from 667 to 671 cm^{-1} (± 0.0023) Week Number						
		0	8	16	24	32	40	48
Blank	23	0.01	NM[a]	NM	0.01	NM	0.01	0.01
N40	23	0.01	NM	NM	0.01	NM	0.01	0.01
Ti	23	0.01	NM	NM	0.01	NM	0.01	0.01
C4130	23	0.02	0.01	0.01	0.01	0.01	NM	NM
I625	23	0.01	NM	0.01	0.01	0.01	0.01	0.01
Blank	150	0.01	0.06	NM	0.08	0.09	0.09	0.09
N40	150	0.01	0.08	NM	0.10	0.12	0.12	0.12
Ti	150	0.01	0.06	NM	0.07	0.08	0.08	0.08
C4130	150	0.01	0.06	0.08	0.09	0.09	0.10	NM
I625	150	0.02	0.04	0.06	0.07	0.07	0.08	0.09

[a] Not measured

samples aged at elevated temperatures. The summary data presented in Tables 16 and 17 indicate the fluorocarbon agents are stable at elevated temperatures for as long as 48 weeks.

7.3.5 Iodotrifluoromethane (CF_3I). This agent contains C-F bonds and a C-I bond. The C-I bond in this molecule is very weak, about 223 kJ/mole (Felder, 1992). This suggests that degradation, particularly at elevated temperatures, may be a factor. Additionally, the presence of moisture in the storage vessels at elevated temperatures was investigated, since water was hypothesized to be an accelerator of degradation. Finally, the possibility that the addition of copper to the cylinders might inhibit degradation was investigated. Therefore, this agent was tested not only at three temperatures, but also at four different conditions of copper and moisture content (Table 1).

The initial spectroscopic analyses of the three lots of CF_3I that were tested indicated that they contained CO_2 and CF_3H. These could have been left from the synthesis or they may be an indication that some degradation of the agent has already occurred. The amount of these impurities depended in large part to the amount of agent still remaining in the storage container when the test cylinders were filled. Since the boiling points of both CO_2 and CF_3H are lower than that of CF_3I, the vapor that is removed from a storage tank is richer in these impurities. Table 10 shows the starting impurity levels of CO_2 and CF_3H for each of the lots used. Notice in the footnotes the amount of agent remaining in the storage tanks when they were analyzed. Lot number 226941712 was analyzed before any of the agent had been used and the CO_2 and CF_3H were the highest. Lot number 226941891 was nearly empty when it was analyzed and it contained little CO_2 and undetectable CF_3H (probably present, but below the detection limit of the method.) In the forthcoming graphical presentations of the data for these impurities, the initial areas vary up and down the y-axis, generally, as a function of which lot of CF_3I was used and when the cylinder was filled (higher areas as a result of fuller tanks.)

7. AGENT STABILITY UNDER STORAGE

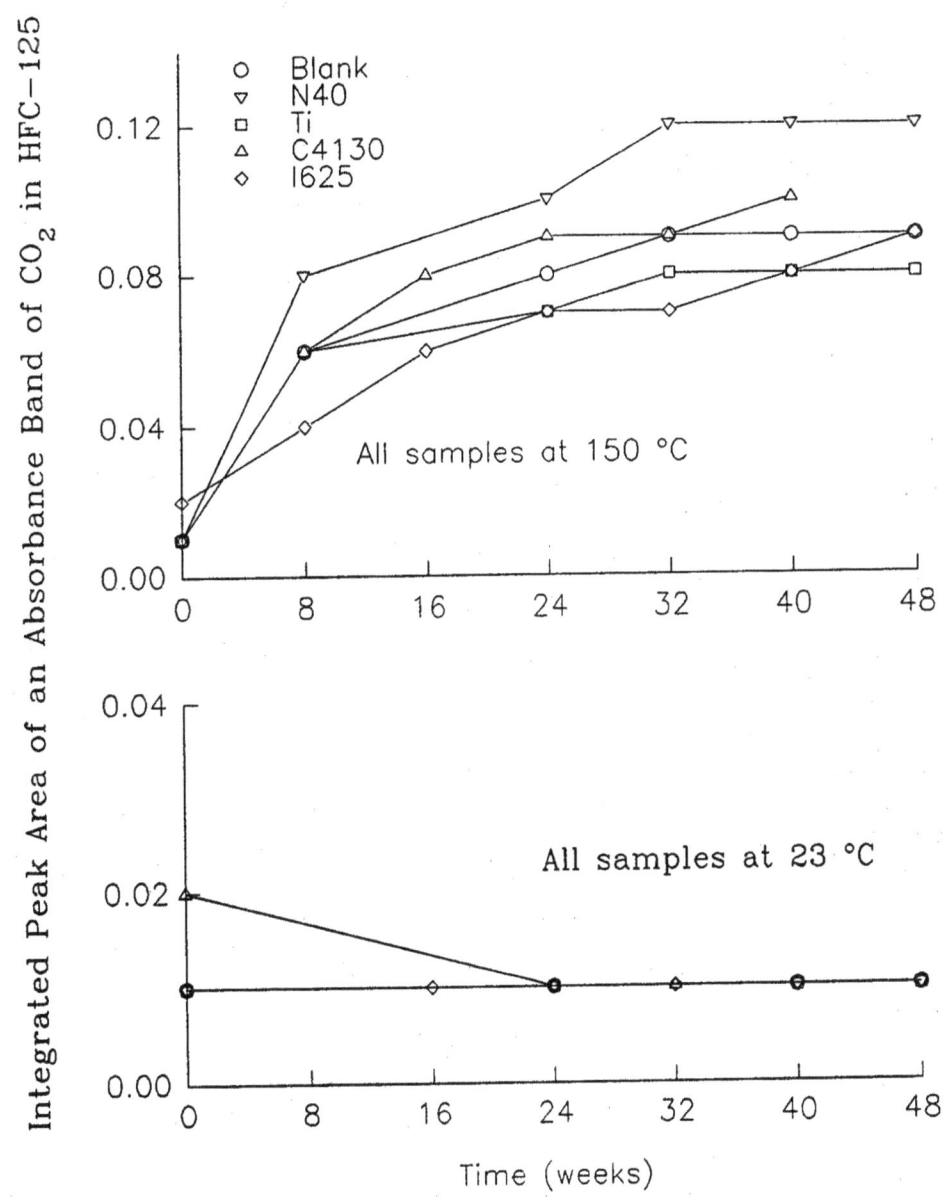

Figure 16. Integrated peak areas for the CO_2 absorbance band at 670 cm^{-1} for all samples of HFC-125 at 23 °C and 150 °C plotted as a function of time.

284   7. AGENT STABILITY UNDER STORAGE

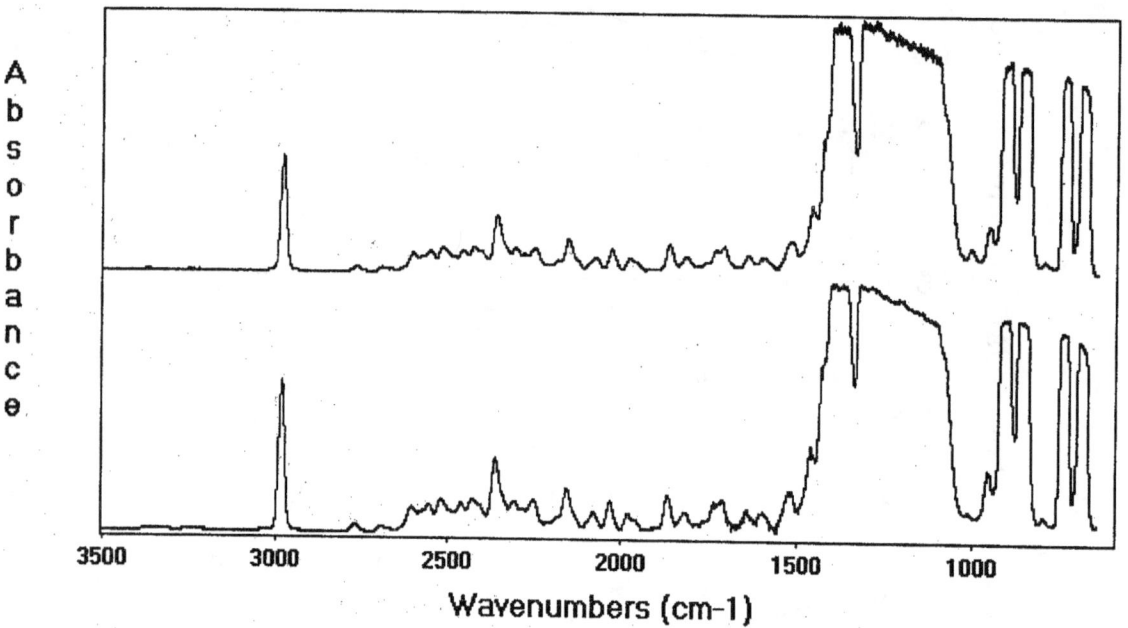

Figure 17. Initial (lower) and 48 week (upper) 5330 Pa spectra of the blank in HFC-227ea at 23 °C.

7. AGENT STABILITY UNDER STORAGE

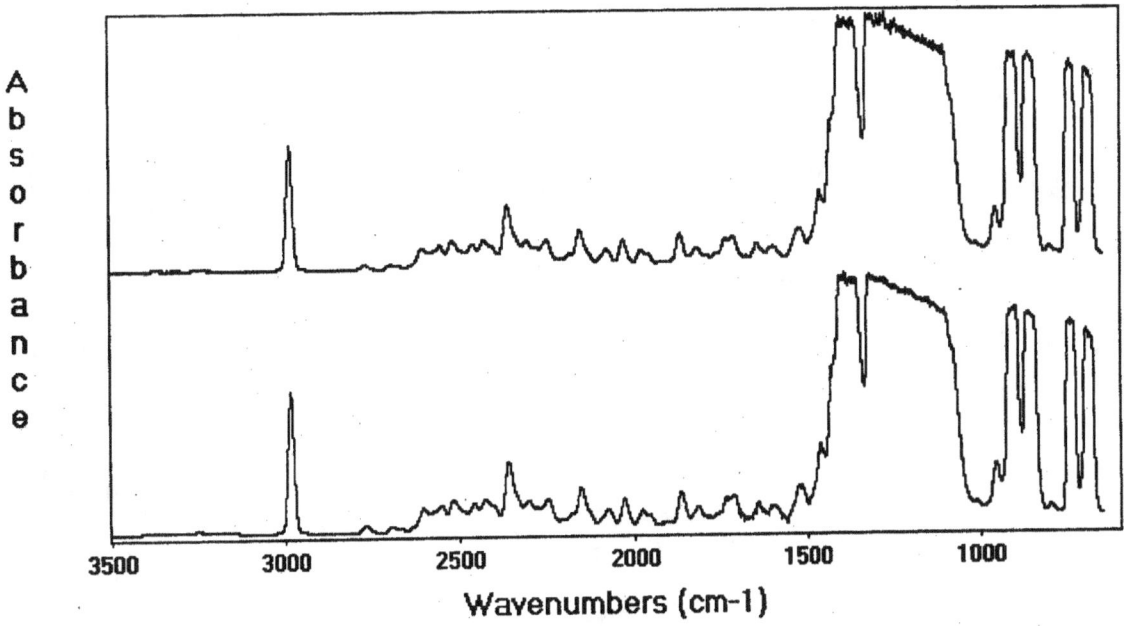

Figure 18. Initial (lower) and 40 week (upper) 5330 Pa spectra for the blank in HFC-227ea at 125 °C.

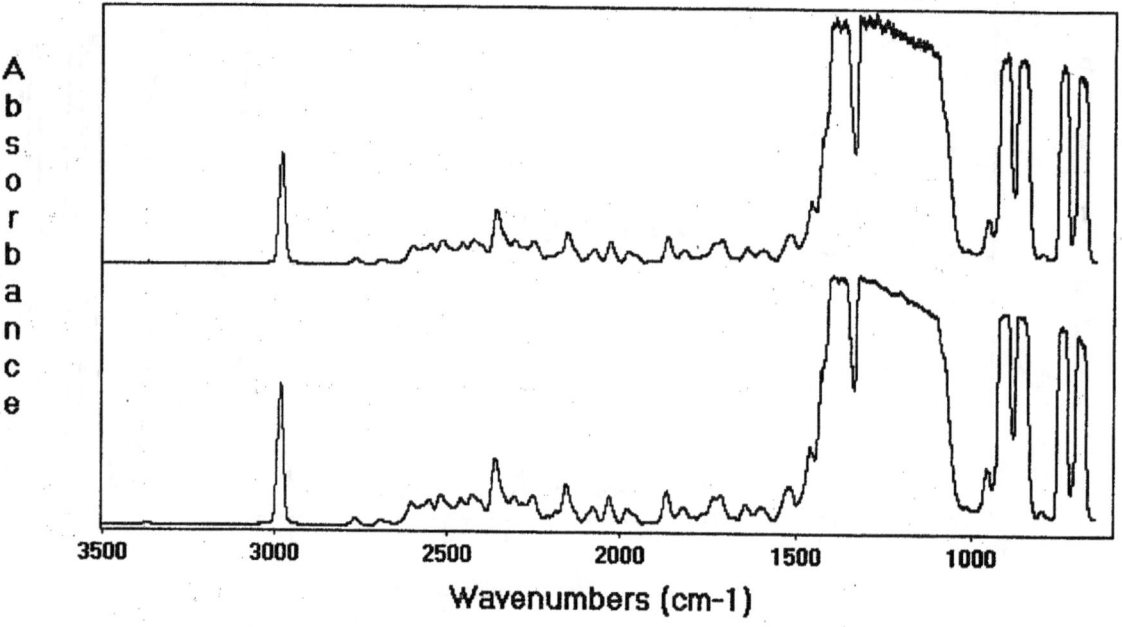

Figure 19. Initial (lower) and 48 week (upper) 5330 Pa spectra of the blank in HFC-227ea at 150 °C.

7. AGENT STABILITY UNDER STORAGE

Table 15. Integrated peak areas for the C-H stretch absorbance band in the 53.3 Pa spectra of HFC-227ea at 23, 125, and 150 °C and the spectral comparisons

		Integrated area from 2966 to 3006 cm^{-1} (± 0.0095)							Spectral comparison (± 0.00044)[a]
		Week number							
Metal	Temp. (°C)	0	8	16	24	32	40	48	
Blank	23	0.22	NM[b]	NM	0.21	NM	0.20	0.21	0.997
N40	23	0.21	NM	NM	0.20	NM	0.19	0.20	0.997
Ti	23	0.21	NM	NM	0.20	NM	0.20	0.20	0.997
C4130	23	0.19	0.19	0.19	0.18	0.20	NM	NM	0.999
I625	23	0.18	NM	0.19	0.18	0.19	0.19	0.19	0.999
Blank	125	0.21	0.22	0.21	0.21	0.23	0.20	NM	0.999
N40	125	0.19	0.20	0.20	0.20	0.21	0.18	NM	0.999
Ti	125	0.19	0.19	0.19	0.19	0.20	0.18	NM	0.999
C4130	125	0.20	0.20	0.19	0.20	0.21	0.20	NM	0.999
I625	125	0.20	0.20	0.20	0.20	0.20	0.20	NM	0.999
Blank	150	0.20	0.20	NM	0.20	0.19	0.19	0.19	0.997
N40	150	0.20	0.20	NM	0.21	0.18	0.20	0.19	0.997
Ti	150	0.23	0.23	NM	0.23	0.21	0.23	0.22	0.997
C4130	150	0.20	0.19	0.19	0.19	0.21	0.20	NM	0.999
I625	150	0.10[c]	0.11	0.12	0.10	0.11	0.10	0.11	0.999

[a] Number of observations was three
[b] Not measured
[c] Improper filling of cylinder at outset of test.

288   7. AGENT STABILITY UNDER STORAGE

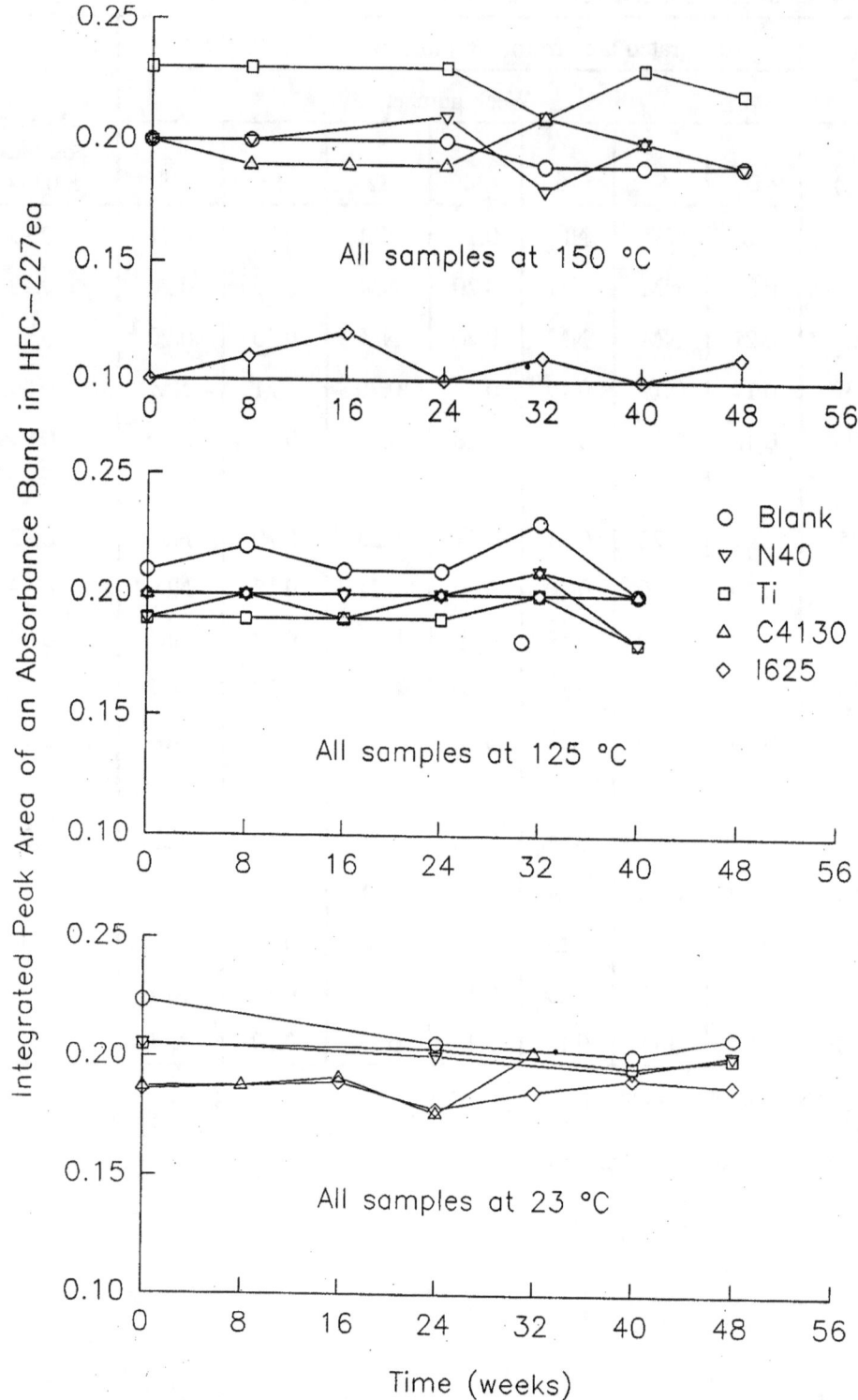

Figure 20. Integrated peak areas for all samples of HFC-227ea at 23 °C, 125 °C, and 150 °C plotted as a function of time.

7. AGENT STABILITY UNDER STORAGE

Table 16. Summary of the initial and final peak areas for the fluorocarbon agents at all temperatures

Metal	Number of weeks	Temp (°C)	Initial	Final	Significant @ 3x uncertainty
FC-218:			area ± (0.15)		
Blank	48	150	6.2	6.5	no
N40	48	150	6.6	7.0	no
Ti	48	150	6.6	6.7	no
C4130	40	150	5.7[a]	6.5	--
I625	48	150	6.5	6.9	no
HFC-125:			area ± (0.042)		
Blank	48	23	2.8	2.7	no
N40	48	23	2.7	2.7	no
Ti	48	23	2.8	2.8	no
C4130	32	23	2.8	2.8	no
I625	48	23	2.8	2.9	no
Blank	48	150	2.8	2.8	no
N40	48	150	2.8	2.8	no
Ti	48	150	2.8	2.8	no
C4130	40	150	2.8	2.9	no
I625	48	150	2.8	2.9	no
HFC-227ea:			area ± (0.0095)		
Blank	48	23	0.22	0.21	no
N40	48	23	0.21	0.20	no
Ti	40	23	0.21	0.20	no
C4130	32	23	0.19	0.20	no
I625	48	23	0.18	0.19	no
Blank	40	125	0.21	0.20	no
N40	40	125	0.19	0.18	no
Ti	40	125	0.19	0.18	no
C4130	40	125	0.20	0.20	no
I625	40	125	0.20	0.20	no
Blank	48	150	0.20	0.19	no
N40	48	150	0.20	0.19	no
Ti	48	150	0.23	0.22	no
C4130	40	150	0.20	0.20	no
I625	48	150	0.10[b]	0.11	no

[a] Improperly filled gas cell during analysis
[b] Cylinder did not fill completely at outset of test.

Table 17. Spectral comparisons of the initial and aged fluorocarbon agents

Agent		FC-218	HFC-125 (± 0.00095)		HFC-227ea (± 0.00044)		
Temp. (°C)		150	23	150	23	125	150
Metal	Weeks tested						
Blank	48	0.996	0.996	0.995	0.997	0.997	0.997
N40	48	0.996	0.997	0.995	0.997	0.997	0.997
Ti	48	0.996	0.996	0.996	0.997	0.997	0.997
C4130	40	0.999	0.998	0.998	0.999	0.999	0.999
I625	48	0.998	0.998	0.998	0.999	0.999	0.999

As mentioned earlier, an absorbance band in the 5330 Pa spectra of CF_3I tested at 100 and 150 °C began to appear around 950 cm^{-1} after 4 weeks of aging. This absorbance was looked for in reference spectra of C_2F_6, COF_2, and CH_3F, but does not appear in any of these compounds. This absorbance band is in the region of C=C bond stretch frequencies. The absorbance band has not been definitely assigned, but may correspond to the presence of ethylene or a fluorinated alkene. Since the corresponding C-H stretch band for ethylene is absent (or below the detection limit at this concentration), it is suspected that the absorbance band might correspond to a fluorinated alkene. Reference spectra for double bonded fluorinated compounds such as C_2F_4, $C_2H_2F_2$ (the 1,1 compound), and C_2H_3F did not have an absorbance band in this region of the spectrum. Not all of the spectra for the fluorinated alkenes are present in our library, so positive identification still remains unresolved. This fluorinated alkene peak and the increase in the impurity peaks are shown in Figure 21. The lower spectrum is the initial, 5330 Pa spectrum of a moist sample without copper and containing phosphate-treated C4130. This is one of the samples that initially had no detectable CF_3H or CO_2. The upper spectrum is that of the corresponding aged sample after 20 weeks at 150 °C. The small peak at 670 cm^{-1} is an absorbance band of CO_2, the peaks at 700, 1400, and 3000 cm^{-1} are from CF_3H, and the absorbance band at 950 cm^{-1} for the fluorinated alkene. This spectrum illustrates quite well the changes that occurred in the CF_3I at elevated temperatures.

The spectrum in Figure 22 shows a spectrum that results from the subtraction of the spectra shown in Figure 21. This subtracted spectrum not only shows the peaks mentioned in Figure 21, but also shows the characteristic peak at 2150 cm^{-1} for carbon monoxide and the other characteristic peak at 2360 cm^{-1} for CO_2. The peaks in the 1200 cm^{-1} region correspond to agent peaks that did not completely subtract out because of their high intensity.

The following data for CF_3I are presented for the dry and moist conditions and with or without the presence of copper. Changes in peak areas for agent, CO_2, CF_3H, and fluorinated alkene are the order of presentation for each test condition.

7.3.5.1 Changes in the Agent Peak Area of CF_3I in the 53.3 Pa Spectra for the Different Conditions. Tables 18-21 show the integrated areas for the absorbance band in the agent and the spectral comparisons. The data presented in these tables are an average of the three replicate gas cell fillings. Figures 23-26 show the respective graphical representations of the data. Figure 23 shows the

7. AGENT STABILITY UNDER STORAGE

data for the dry condition without copper. At 23 °C, with the exception of the outlier for Nitronic 40 (a result of improper filling of the gas cell), the data are closely bunched and remain relatively constant. The same was true for the data at 100 °C. The data for 150 °C were more spread out but generally seem to equilibrate after a slight decrease.

Figure 24 shows the peak area data for the dry condition with copper. At 23 °C, the areas tended to drift downward slightly, then equilibrated. Overall, there is not much change in the data for any of the metals over the course of the study. The blank at 100 °C was scattered but also unchanged. The areas at 150 °C were drifting downward consistently. This suggests that copper is not inhibiting the degradation of CF_3I at the elevated temperature.

Figure 25 shows the peak area data for the moist condition without copper. There were no 23 °C controls for this condition. The areas for the blank at 100 °C remained fairly constant over the study. The peak areas at 150 °C drifted downward slightly, but reached a minimum at about 32 weeks and started to equilibrate. All of the H_2O may have been consumed by this time. The areas at 150 °C were more scattered than the previous areas at 150 °C for the dry condition.

Figure 26 shows the peak area data for the moist condition with copper. Once again there were no controls at 23 °C for this condition. The areas were changing similarly to the moist condition without copper shown in Figure 25.

The above graphs suggest that small decreases in agent peak area were occurring, especially at 150 °C and in the moist condition more so than the dry condition. It appears that copper is not inhibiting the degradation, H_2O causes accelerated degradation, and both combined cause about the same amount, all at 150 °C. The other metals seem to have no effect. Since the samples at 100 °C showed no effect, it may be that at ≤ 100 °C there is little loss of CF_3I.

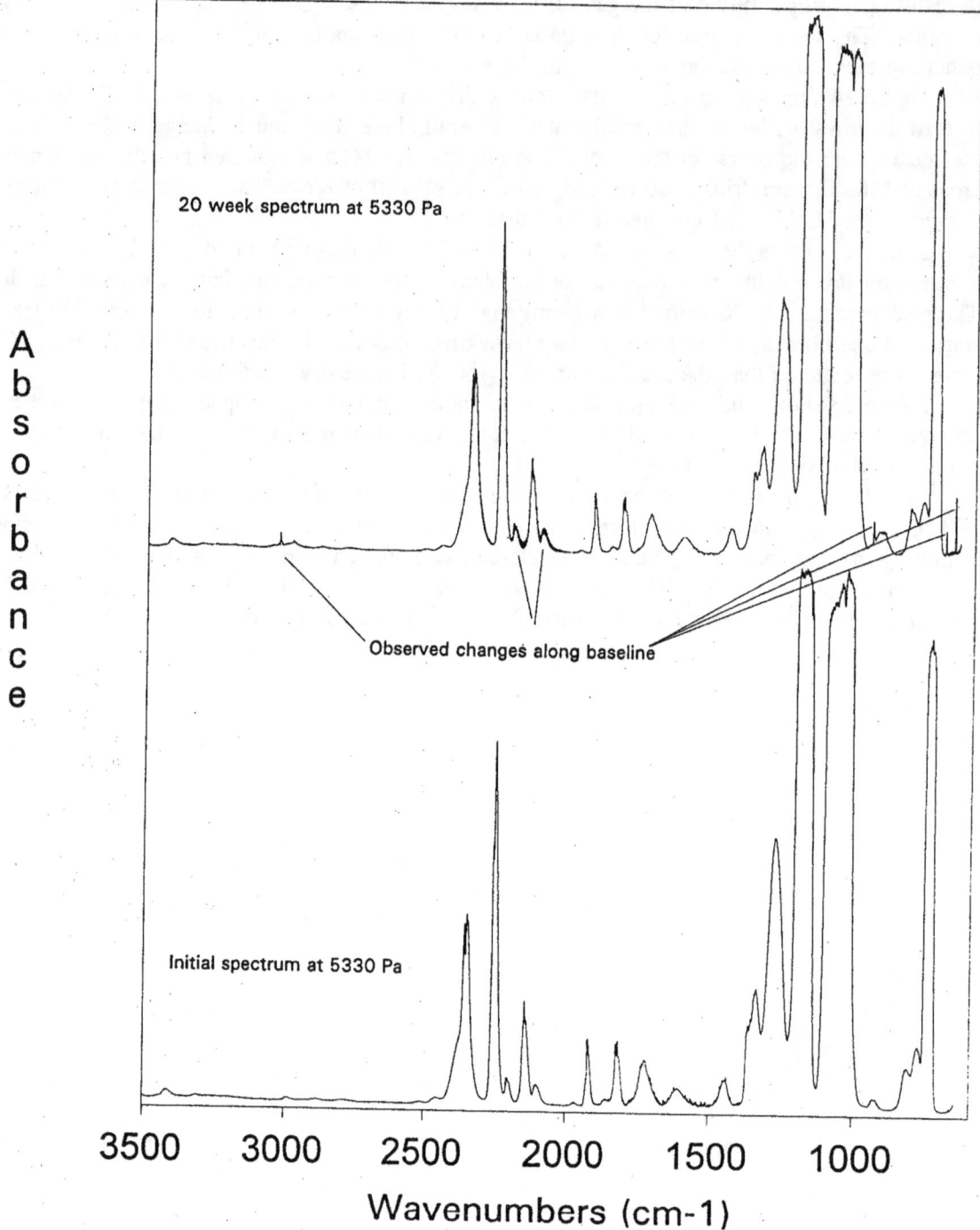

Figure 21. Illustration of the changes taking place in CF$_3$I samples at elevated temperatures.

7. AGENT STABILITY UNDER STORAGE

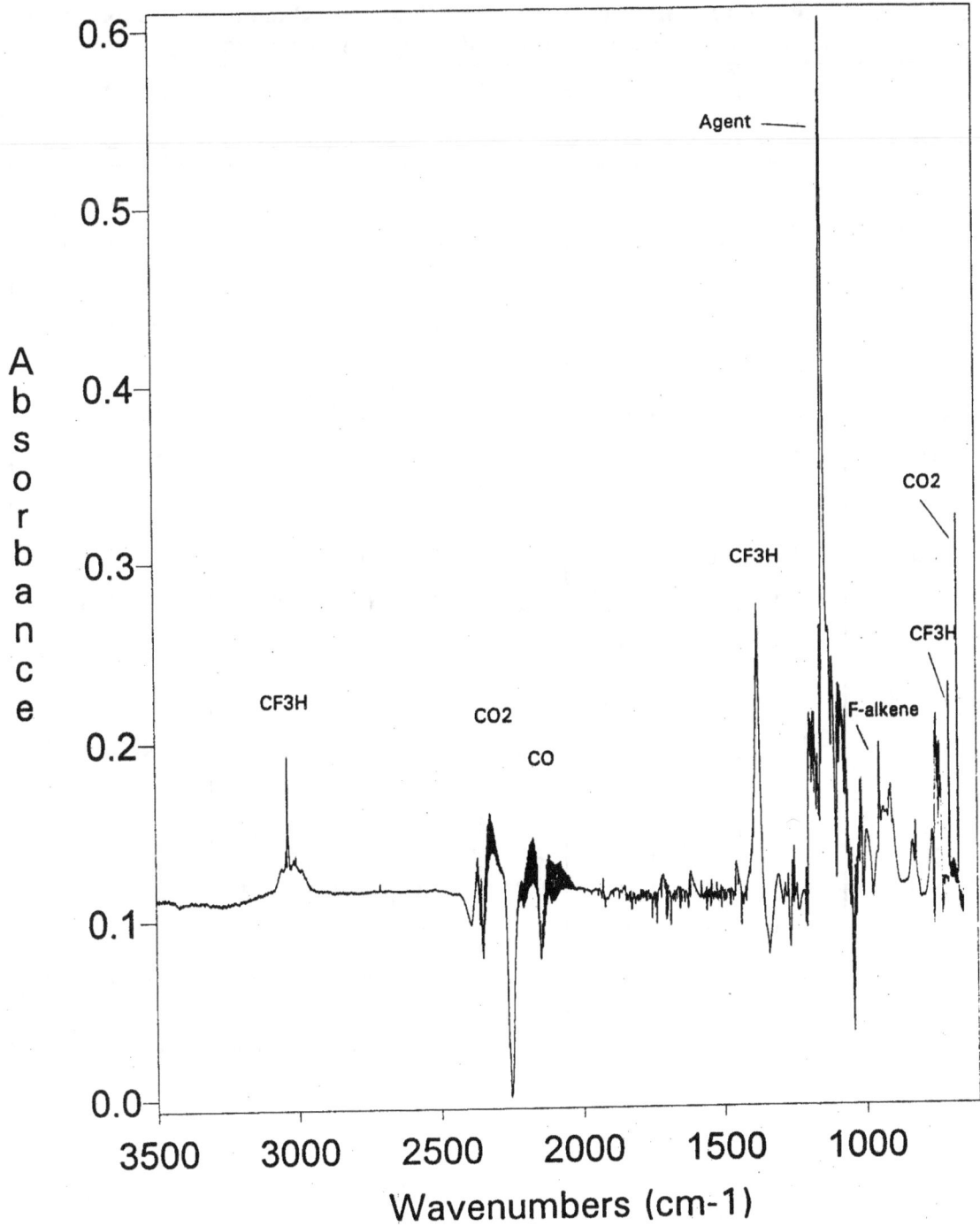

Figure 22. The subtracted spectrum at 5330 Pa for CF_3I in the presence of C4130 alloy and moisture at 150 °C after 20 weeks.

Table 18. Integrated peak areas for an absorbance band of CF_3I in the 53.3 Pa spectra tested in the dry condition without copper and the spectral comparisons

Metal	Temp. (°C)	Integrated area from 2219 to 2274 cm^{-1} (± 0.011)								
		Week number								
		0	4	8	16	20	24	28	32	36
Blank	23	0.25	0.25	0.24	0.24	0.25	0.22	NM[a]	0.24	NM
N40	23	0.19	0.27	0.25	0.25	0.25	0.24	NM	0.26	NM
Ti	23	0.26	0.25	0.25	0.25	0.23	0.23	NM	0.24	NM
C4130	23	0.24	0.25	0.26	0.24	0.22	0.24	NM	0.23	NM
I625	23	0.27	0.26	0.24	0.24	0.24	0.24	NM	0.24	0.24
Blank	100	0.25	0.24	0.25	0.25	0.24	0.25	0.24	0.25	0.21
N40	100	0.24	0.25	0.25	0.24	0.24	0.24	0.23	0.25	--[b]
Ti	100	0.25	0.25	0.24	0.24	0.24	0.23	0.24	0.25	0.24
C4130	100	0.25	0.25	0.24	0.25	0.24	0.23	NM	0.25	0.24
I625	100	0.25	0.26	0.25	0.25	0.25	0.22	NM	0.23	0.24
Blank	150	0.28	0.28	0.28	0.27	0.26	0.25	0.25	0.25	0.25
N40	150	0.27	0.26	0.27	0.25	0.25	0.24	0.24	0.23	0.24
Ti	150	0.28	0.27	0.27	0.26	0.25	0.25	0.26	0.25	0.25
C4130	150	0.29[c]	0.24	0.24	0.24	0.26	0.21	NM	0.21	0.23
I625	150	0.25	0.25	0.24	0.23	0.22	0.23	NM	0.22	0.22

7. AGENT STABILITY UNDER STORAGE

Table 18 (cont.). Integrated peak areas for an absorbance band of CF_3I in the 53.3 Pa spectra tested in the dry condition without copper and the spectral comparisons

Metal	Temp. (°C)	Integrated area from 2219 to 2274 cm^{-1} (± 0.011)				Spectral comparison (0.0018)[d]
		Week number				
		40	44	48	52	
Blank	23	0.23	NM	0.23	NM	0.996
N40	23	0.25	NM	0.24	NM	0.998
Ti	23	0.24	NM	0.23	NM	0.998
C4130	23	0.25	NM	NM	NM	0.999
I625	23	0.25	0.24	NM	NM	0.998
Blank	100	0.25	0.25	NM	NM	0.999
N40	100	--[b]	--[b]	--[b]	--[b]	0.997
Ti	100	0.24	0.24	NM	NM	0.998
C4130	100	0.24	NM	NM	NM	0.999
I625	100	0.24	NM	NM	NM	0.999
Blank	150	0.25	0.25	0.24	0.25	0.991
N40	150	0.24	0.23	0.23	0.25	0.995
Ti	150	0.26	0.25	0.25	0.25	0.996
C4130	150	0.23	NM	NM	NM	0.994
I625	150	0.22	0.22	NM	NM	0.988

[a] Not measured
[b] Cylinder emptied after 32 weeks as a result of leak
[b] Suspected to be an outlier because of improper gas cell filling
[c] Number of observations was eight.

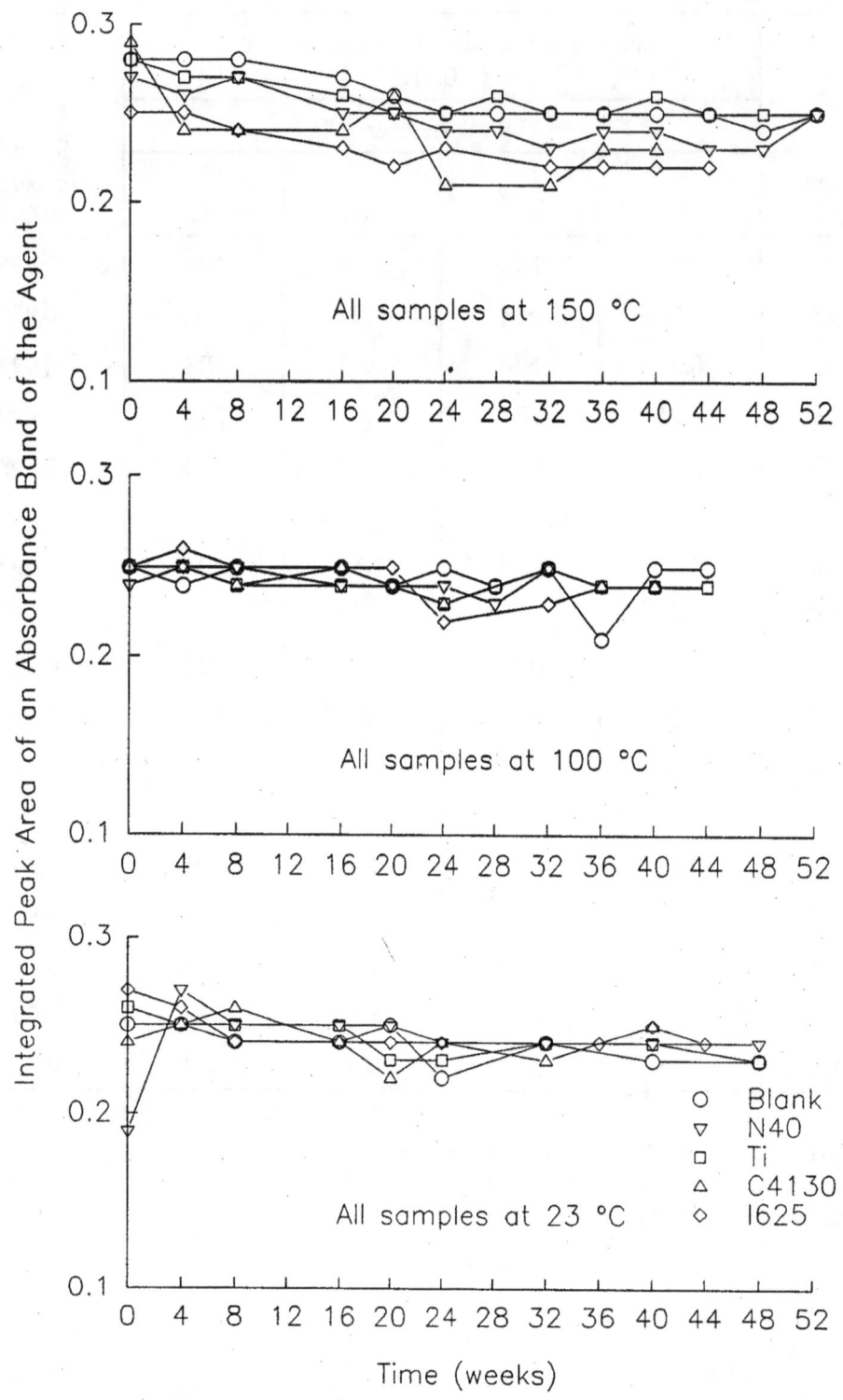

Figure 23. Integrated agent peak areas for all samples of CF_3I tested in the dry condition without copper at 23 °C, 100 °C, and 150 °C plotted as a function of time.

7. AGENT STABILITY UNDER STORAGE

Table 19. Integrated peak areas for an absorbance band of CF_3I in the 53.3 Pa spectra tested in the dry condition with copper and the spectral comparisons

		Integrated area from 2219 to 2274 cm^{-1} (±0.011)								
		Week number								
Metal	Temp. (°C)	0	4	8	16	20	24	28	32	36
CDA 110 Blank	23	0.26	0.25	0.24	0.25	0.23	0.24	NM[a]	0.24	NM
CDA 110 N40	23	0.26	0.25	0.24	0.23	0.23	0.24	NM	0.24	NM
CDA 110 Ti	23	0.27	0.26	0.25	0.24	0.24	0.24	NM	0.24	NM
CDA 110 C4130	23	0.23	0.25	0.24	0.23	0.23	0.24	NM	0.21	NM
CDA 110 I625	23	0.26	0.25	0.25	0.23	0.24	0.23	NM	0.24	0.24
CDA 110 Blank	100	0.25	0.25	0.23	0.23	0.25	0.23	0.25	0.26	--[b]
CDA 110 Blank	150	0.29	0.26	0.26	0.26	0.25	0.24	0.25	0.23	0.24
CDA 110 N40	150	0.26	0.27	0.27	0.25	0.24	0.23	0.23	0.23	0.23
CDA 110 Ti	150	0.28	0.26	0.27	0.26	0.24	0.24	0.24	0.24	0.24
CDA 110 C4130	150	0.24	0.25	0.25	0.25	0.23	0.22	NM	0.23	0.25
CDA 110 I625	150	0.25	0.24	0.25	0.22	0.22	0.23	NM	0.23	0.23

Table 19 (cont.). Integrated peak areas for an absorbance band of CF_3I in the 53.3 Pa spectra tested in the dry condition with copper and the spectral comparisons

Metal	Temp. (°C)	Integrated area from 2219 to 2274 cm^{-1} (±0.011) Week number				Spectral comparison (±0.0018)[c]
		40	44	48	52	
CDA 110 Blank	23	0.25	NM	0.23	NM	0.997
CDA 110 N40	23	0.23	NM	0.24	NM	0.997
CDA 110 Ti	23	0.24	NM	0.25	NM	0.997
CDA 110 C4130	23	0.24	NM	NM	NM	0.999
CDA 110 I625	23	0.24	0.23	NM	NM	0.999
CDA 110 Blank	100	--[b]	--[b]	--[b]	--[b]	0.999
CDA 110 Blank	150	0.24	0.23	0.22	0.23	0.980
CDA 110 N40	150	0.21	0.22	0.20	0.19	0.875
CDA 110 Ti	150	0.23	0.24	0.23	0.23	0.985
CDA 110 C4130	150	0.22	NM	NM	NM	0.997
CDA 110 I625	150	0.22	0.23	NM	NM	0.994

[a] Not measured
[b] Cylinder emptied after 32 weeks as a result of leak
[c] Number of measurements was eight.

7. AGENT STABILITY UNDER STORAGE

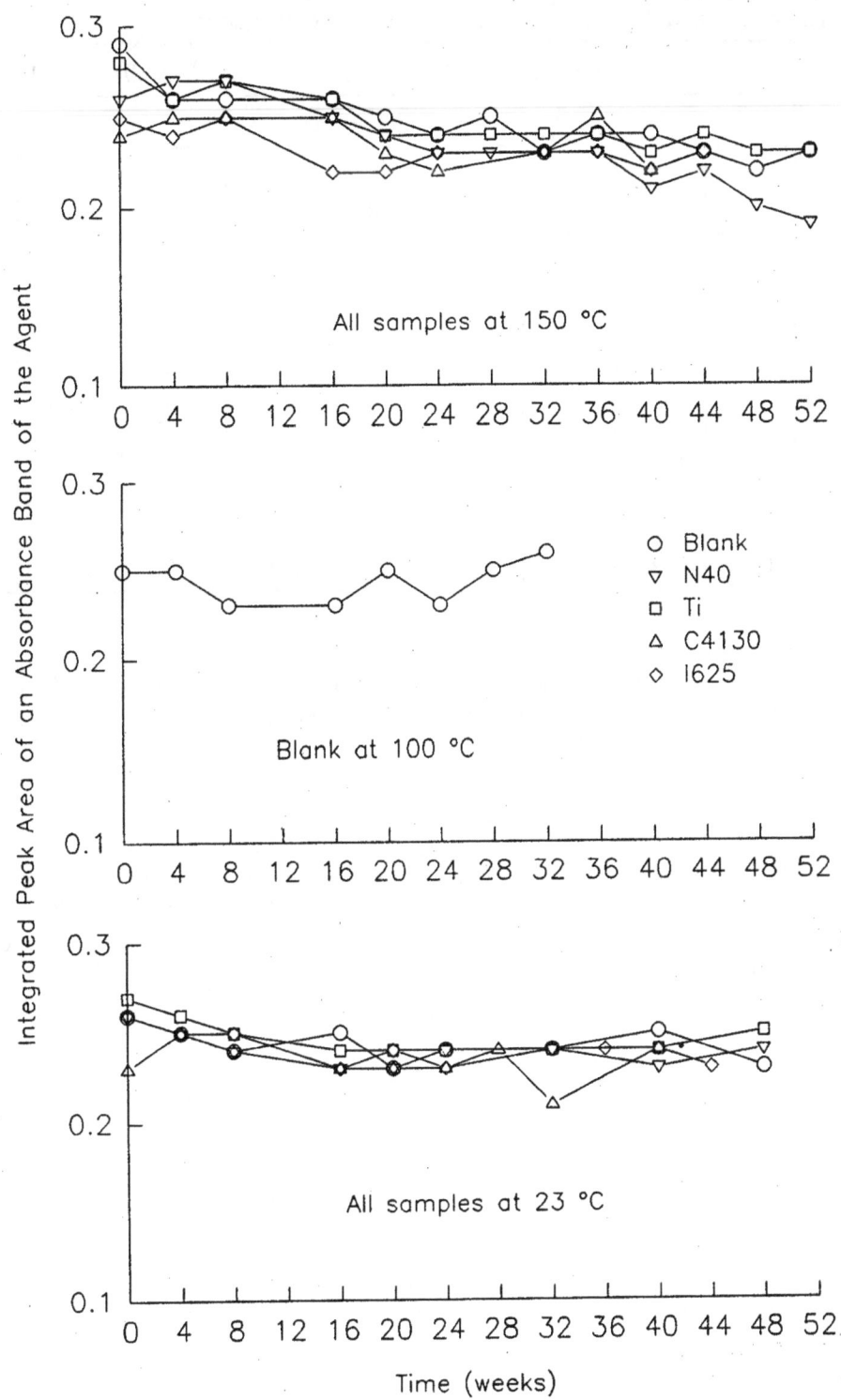

Figure 24. Integrated agent peak areas for all samples of CF$_3$I tested in the dry condition with copper at 23 °C, 100 °C, and 150 °C plotted as a function of time.

Table 20. Integrated peak areas for an absorbance band in the 53.3 Pa spectra of CF_3I tested in the moist condition without copper and the spectral comparisons

Metal	Temp. (°C)	Integrated area from 2219 to 2274 cm^{-1} (±0.011)								
		Week number								
		0	4	8	16	20	24	28	32	36
Blank	100	0.25	0.24	0.24	0.24	0.23	0.23	0.24	0.22	0.23
Blank	150	0.27	0.25	0.27	0.25	0.24	0.24	0.24	0.24	0.23
N40	150	0.27	0.25	0.27	0.25	0.24	0.24	0.25	0.24	0.20
Ti	150	0.28	0.25	0.27	0.27	0.24	0.24	0.24	0.24	0.24
C4130	150	0.24	0.25	0.23	0.23	0.22	0.21	NM[a]	0.20	0.22
I625	150	0.25	0.24	0.24	0.22	0.22	0.22	NM	0.21	0.21

Table 20 (cont.). Integrated peak areas for an absorbance band in the 53.3 Pa spectra of CF_3I tested in the moist condition without copper and the spectral comparisons

Metal	Temp. (°C)	Integrated area from 2219 to 2274 cm^{-1} (±0.011)				Spectral Comparison (±0.0018)[b]
		Week number				
		40	44	48	52	
Blank	100	0.23	0.24	NM	NM	0.999
Blank	150	0.24	0.23	0.24	0.23	0.984
N40	150	0.25	0.25	0.25	0.23	0.992
Ti	150	0.24	0.24	0.23	0.23	0.990
C4130	150	0.21	NM	NM	NM	0.990
I625	150	0.22	0.21	NM	NM	0.983

[a] Not measured
[b] Number of observations was eight.

7. AGENT STABILITY UNDER STORAGE

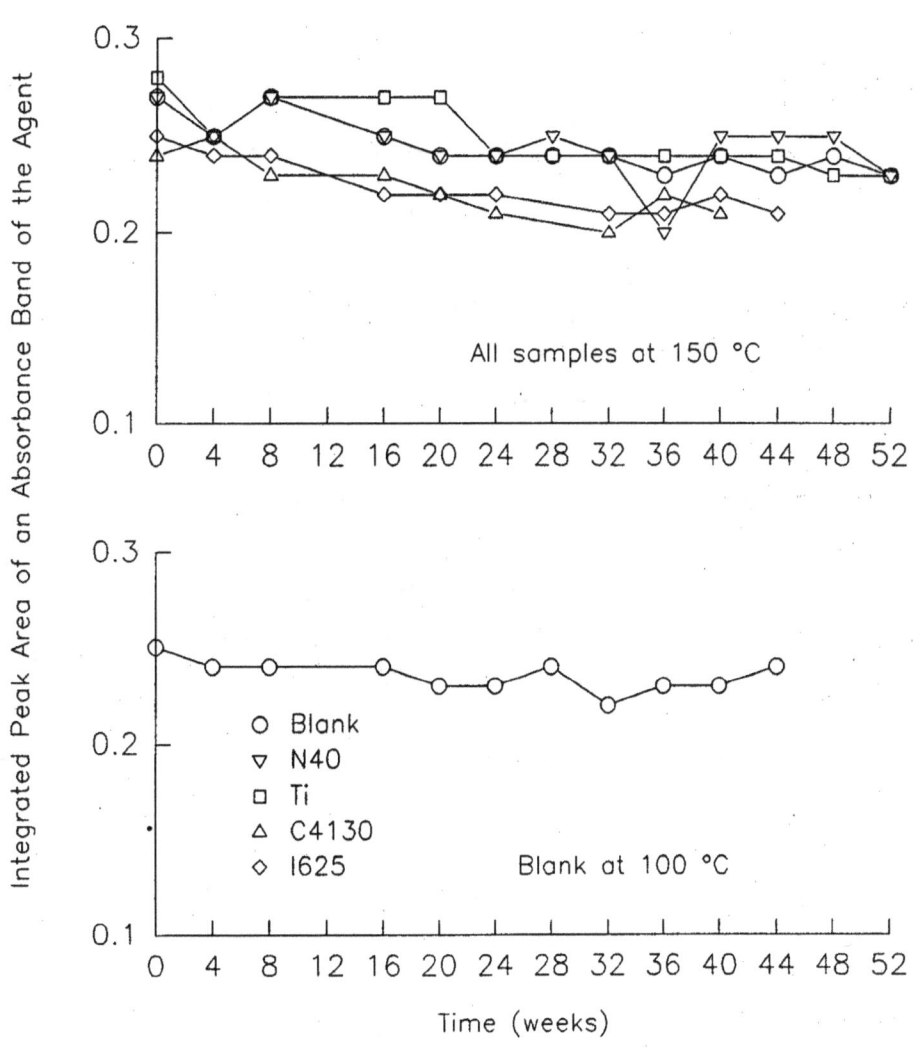

Figure 25. Integrated agent peak areas for all samples of CF_3I tested in the moist condition without copper at 100 °C and 150 °C plotted as a function of time.

Table 21. Integrated peak areas for an absorbance band in the 53.3 Pa spectra of CF_3I tested in the moist condition without copper and the spectral comparisons

Metal	Temp. (°C)	Integrated area from 2219 to 2274 cm^{-1} (±0.011)								
		Week number								
		0	4	8	16	20	24	28	32	36
CDA 110 Blank	100	0.25	0.24	0.23	0.24	0.22	0.22	0.23	0.23	0.23
CDA 110 Blank	150	0.27	0.25	0.26	0.25	0.25	0.22	0.23	0.22	0.22
CDA 110 N40	150	0.27	0.25	0.26	0.24	0.25	0.21	0.22	0.22	0.21
CDA 110 Ti	150	0.28	0.26	0.27	0.25	0.26	0.23	0.23	0.23	0.24
CDA 110 C4130	150	0.23	0.24	0.22	0.23	0.22	0.23	NM[a]	0.21	0.23
CDA 110 I625	150	0.23	0.23	0.23	0.23	0.21	0.22	NM	0.22	0.22

Table 21 (cont.). Integrated peak areas for an absorbance band in the 53.3 Pa spectra of CF_3I tested in the moist condition without copper and the spectral comparisons

Metal	Temp. (°C)	Integrated area from 2219 to 2274 cm^{-1} (±0.011)				Spectral comparison (±0.0018)[a]
		Week number				
		40	44	48	52	
CDA 110 Blank	100	0.23	0.24	NM	NM	0.998
CDA 110 Blank	150	0.23	0.22	0.22	0.22	0.981
CDA 110 N40	150	0.21	0.21	0.20	0.19	0.949
CDA 110 Ti	150	0.23	0.23	0.22	0.24	0.990
CDA 110 C4130	150	0.21	NM	NM	NM	0.998
CDA 110 I625	150	0.21	NM	NM	NM	0.994

[a] Not measured
[b] Number of observations was eight.

7. AGENT STABILITY UNDER STORAGE

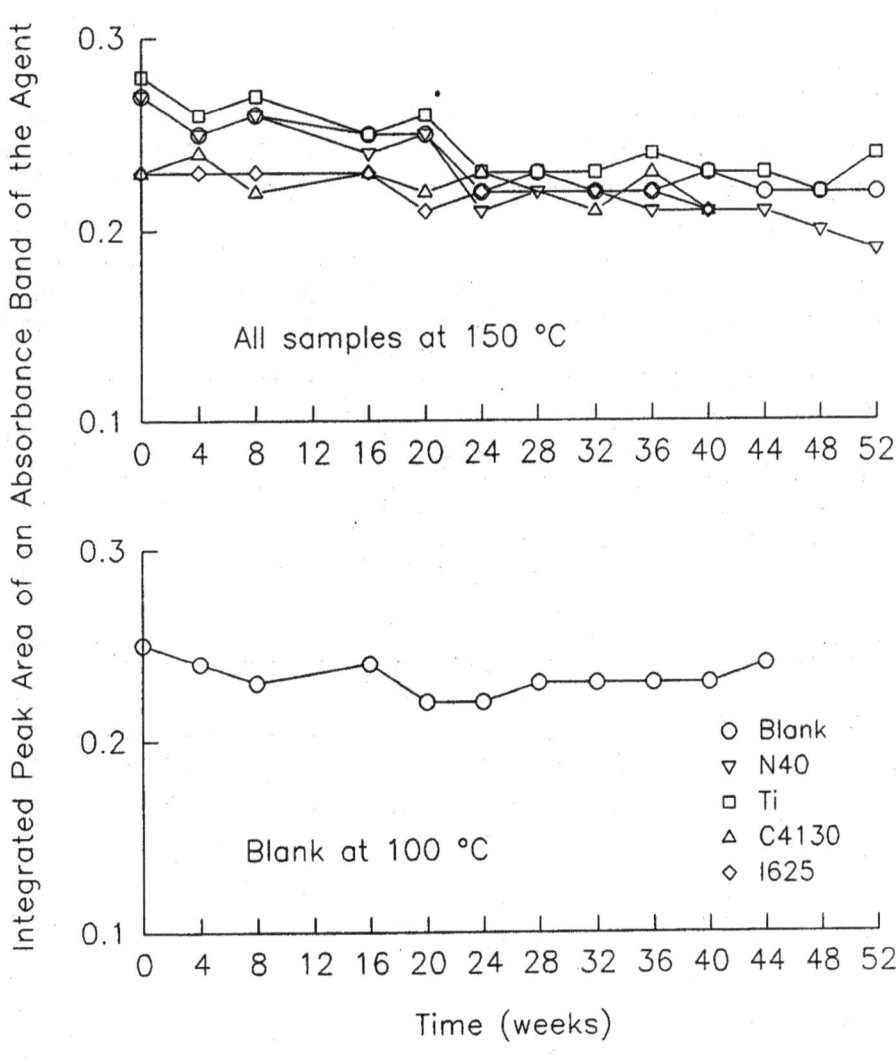

Figure 26. Integrated agent peak areas for all samples of CF_3I tested in the moist condition with copper at 100 °C and 150 °C plotted as a function of time.

7.3.5.2 Changes in the Impurity Peak Areas Appearing in the 5330 Pa Spectra.

The peak area data in the remaining tables represent a single value for each analysis obtained over the course of the study. Tables 22-25 show the integrated peak areas for the absorbance band at 670 cm^{-1} for CO_2 for all temperatures and test conditions of CF_3I. Figures 27-30 are the respective graphical representations of the data. In both of the test conditions where ambient samples were tested (Figures 27 and 28), no changes in the peak areas were noted. The CO_2 concentration in the initial cylinders ranged from 26 to 460 µL/L (based on a 0.1 % by volume standard of CO_2 in N_2.) However, all samples tested at elevated temperatures showed an increase in the CO_2 peak area. Even samples that initially contained CO_2 as low as 26 µL/L (C4130 and I625) experienced an increase in CO_2 concentration to roughly 900 µL/L. The CO_2 concentration in the cylinder containing I625 metal in the dry condition with and without copper and the moist condition without copper increased from 400 µL/L to nearly 1900 µL/L. The dramatic increase in the CO_2 did not occur for the I625 in the moist condition with copper, but the initial CO_2 content in this cylinder was only 26 µL/L. The CO_2 content in the cylinders with the other metals only increased about half this amount. The CO_2 production continued to increase after 52 weeks. Also, all samples tested at 100 °C showed an increase in the CO_2 peak area. The actual concentration was about half that in the 150 °C cylinders, except for C4130 and I625 in the dry condition without copper, which had very little CO_2 present at the outset of testing.

Tables 26-29 show the integrated peak areas for the absorbance band at 700 cm^{-1} in the 5330 Pa spectrum for the CF_3H impurity. Figures 31-34 show the graphs for the respective data. There was no increase in the CF_3H peak area for any of the conditions at 23 °C or 100 °C. However, at 150 °C the generation of CF_3H was increasing with time and, like the CO_2, was still increasing after 52 weeks. Noticeably different for the CF_3H increase was a longer time before the slope started to show an increase. I625 in both the dry and moist conditions without copper (Figures 31 and 33, respectively) showed a more rapid increase than the other metals. In the dry condition with copper (Figure 32) it was the nitronic 40 metal that showed a more rapid increase.

Tables 30-33 show the integrated peak areas for the absorbance band at 950 cm^{-1} of the fluorinated alkene. Figures 35-38 show the graphs for the respective data. In this case, none of the initial samples showed any absorbance in the 950 cm^{-1} band and at 23 °C for as long as 48 weeks there was no evidence of formation. After about 8 weeks at 100 °C, a small but easily resolved peak started to appear. In both the dry and moist condition without copper (Figures 35 and 37, respectively), the concentration was still increasing after 8 weeks. In both the dry and moist condition with copper (Figures 36 and 38, respectively), the concentration increased then equilibrated. However, after only 4 weeks at 150 °C, the concentration of the fluorinated alkene increased rapidly then began to equilibrate after 20 weeks. In the dry condition without copper at 150 °C (Figure 35), the fluorinated alkene content maximized around 20 weeks, then equilibrated. There may have been some differentiation of the other metals with copper present at 150 °C (Figures 36 and 38). The blank containing only copper strips, and the Ti-15-3-3-3 with copper strips in the dry condition (Figure 36) maximized in about 8 weeks, then started to drop and after 52 weeks was approaching a very low level. The Nitronic 40 and I625 metals with the copper caused more of the fluorinated alkene to appear and the decline was even more gradual. The C4130 in combination with copper resulted in the most fluorinated alkene appearing and there was no decline in the compound.

Figure 38 shows the data for the moist condition with copper. Note the similarity of this data to Figure 36 with respect to the alkene content and metal, but equilibration occurred after 20 weeks. Also of note is that the blank, which contains only copper strips, did not produce detectable amounts of the alkene at 150 °C. However, the blank at 100 °C showed the presence of detectable fluorinated alkene; the peak area did appear to equilibrate with time. A possible explanation is that the copper is causing the double bond to break, and a higher temperature is needed to activate the reaction.

7. AGENT STABILITY UNDER STORAGE

Although the characteristic band at 2150 cm^{-1} for CO appears in all of the elevated temperature samples for CF$_3$I, this absorbance peak was not integrated as a function of time as with the other impurity peaks. It is reasonable to expect the presence of CO along with CO$_2$.

The initial and final spectra for all samples of CF$_3$I are found in appendixes C, D, E, and F. These spectra are included for reference in order of dry condition, without copper (Appendix C); dry condition, with copper (Appendix D); moist condition, without copper (Appendix E) and moist condition, with copper (Appendix F).

7.3.6 Summary of the CF$_3$I Area Data. The peak area data for an absorbance band of the CF$_3$I presented in the above graphs (Figures 23-26) suggest that the areas are decreasing slightly as a function of time at elevated temperatures, especially with moisture present. Table 34 is a summary of the data in Tables 18-21 showing only the initial and final peak areas along with the time of aging at all temperatures and conditions. The last column shows whether the final area was significantly decreased at the 99 % confidence interval. None of the samples at 23 or 100 °C showed a statistically significant change. However, some samples at 150 °C aging started to show statistically significant decreases, especially those that aged for 52 weeks. In the dry condition without copper at 150 °C none of the final areas were significantly decreased (the initial area for this sample was higher than expected and might be an outlier.) In the dry condition with copper, three of the areas decreased significantly and the other two which did not age as long appear to be decreasing and in time will become significantly lower. In the moist condition without copper, four of the five samples show a significant decrease, the other sample appears to be going down, also. Finally, in the moist condition with copper, three of the samples show a significant decrease. The other two samples are decreasing and at longer time might become significant. Even though some of the areas at 150 °C are showing "statistically significant" differences, the actual loss in agent is probably quite small and poses no problem to the fire extinguishing capability of the CF$_3$I.

The data presented in Table 35 show the spectral comparisons for all of the CF$_3$I samples aged at 150 °C compared to their respective samples maintained at 23 °C. Keep in mind that the spectral comparisons for the elevated temperature samples were for the longest aged sample compared to its initial spectrum and represents a single value, whereas the spectral comparisons for the 23 °C samples are an average of each aged spectrum compared to its respective initial spectrum. Although not all samples have been aged the same length of time, these data do suggest some significant changes in the aged spectra with time. Also one should keep in mind that the impurity peaks are small in comparison to the agent peaks, so the increase in the areas of all combined peaks resulted in significant changes. Since the agent concentration in the cylinders is greater than 100 000 µL/L, and the highest measured concentration of CO$_2$ impurity was only 1900 µL/L, the amount of impurity from all sources is probably not enough to have an affect on the overall performance of the CF$_3$I. Some fluorinated alkenes are highly toxic, and if the peak at 950 cm^{-1} is a result of an alkene, the only question is whether enough is being generated to form a harmful mixture.

The data for CF$_3$I suggested that the weak C-I bond in CF$_3$I was indeed breaking at the 100 and 150 °C temperatures. Once the bond broke the radical was free to recombine with another CF$_3$ radical, an H atom or other radicals in the matrix. Since the CF$_3$H concentration in the cylinders continued to increase with time and the CF$_3$I agent peak continued to show slight decreases, the CF$_3$ radical combined with a hydrogen atom to form a more stable CF$_3$H molecule. The presence of moisture in the sample probably provided an increased source of hydrogen atoms, therefore the degradation process was accelerated.

Table 22. Integrated peak areas for an absorbance band of CO_2 in the 5330 Pa spectra of CF_3I in the dry condition without copper

Metal	Temp. (°C)	Integrated area from 667 to 671 cm^{-1} (±0.0048) Week number								
		0	4	8	16	20	24	28	32	36
Blank	23	0.09	0.09	0.09	0.09	0.09	0.09	NM[a]	0.09	NM
N40	23	0.08	0.08	0.08	0.08	0.08	0.07	NM	0.08	NM
Ti	23	--[b]	0.08	0.09	0.08	0.09	0.08	NM	0.08	NM
C4130	23	0.01	0.01	0.01	0.01	0.01	0.01	NM	0.01	NM
I625	23	0.18	0.17	0.16	0.16	0.16	0.17	NM	0.16	0.17
Blank	100	0.14	0.15	0.14	0.15	0.16	0.16	0.17	0.17	0.16
N40	100	0.14	0.14	0.14	0.16	0.16	0.16	0.17	0.17	--[c]
Ti	100	0.14	0.14	0.14	0.15	0.15	0.16	0.17	0.16	0.16
C4130	100	0.01	0.01	0.02	0.02	0.03	0.03	NM	0.04	0.05
I625	100	0.01	0.01	0.01	0.02	0.02	0.02	NM	0.03	0.04
Blank	150	0.18	0.24	0.25	0.27	0.30	0.30	0.31	0.35	0.36
N40	150	0.17	0.26	0.27	0.28	0.30	0.29	0.30	0.31	0.32
Ti	150	0.17	0.20	0.22	0.24	0.28	0.25	0.26	0.29	0.30
C4130	150	0.01	0.09	0.12	0.16	0.20	0.23	NM	0.27	0.32
I625	150	0.17	0.68	0.66	0.69	0.68	0.72	NM	0.69	0.72

7. AGENT STABILITY UNDER STORAGE

Table 22 (cont.). Integrated peak areas for an absorbance band of CO_2 in the 5330 Pa spectra of CF_3I in the dry condition without copper

Metal	Temp. (°C)	Integrated area from 667 to 671 cm^{-1} (± 0.0048)			
		Week number			
		40	44	48	52
Blank	23	0.08	NM	0.09	NM
N40	23	0.07	NM	0.08	NM
Ti	23	0.08	NM	0.09	NM
C4130	23	0.01	NM	NM	NM
I625	23	0.16	0.17	NM	NM
Blank	100	0.18	0.19	NM	NM
N40	100	--c	--c	--c	--c
Ti	100	0.18	0.19	NM	NM
C4130	100	0.05	NM	NM	NM
I625	100	0.04	NM	NM	NM
Blank	150	0.38	0.41	0.39	0.42
N40	150	0.32	0.32	0.31	0.33
Ti	150	0.32	0.34	0.35	0.40
C4130	150	0.36	NM	NM	NM
I625	150	0.73	0.75	NM	NM

[a] Not measured
[b] Lost spectrum file
[c] Cylinder emptied after 32 weeks from a leak.

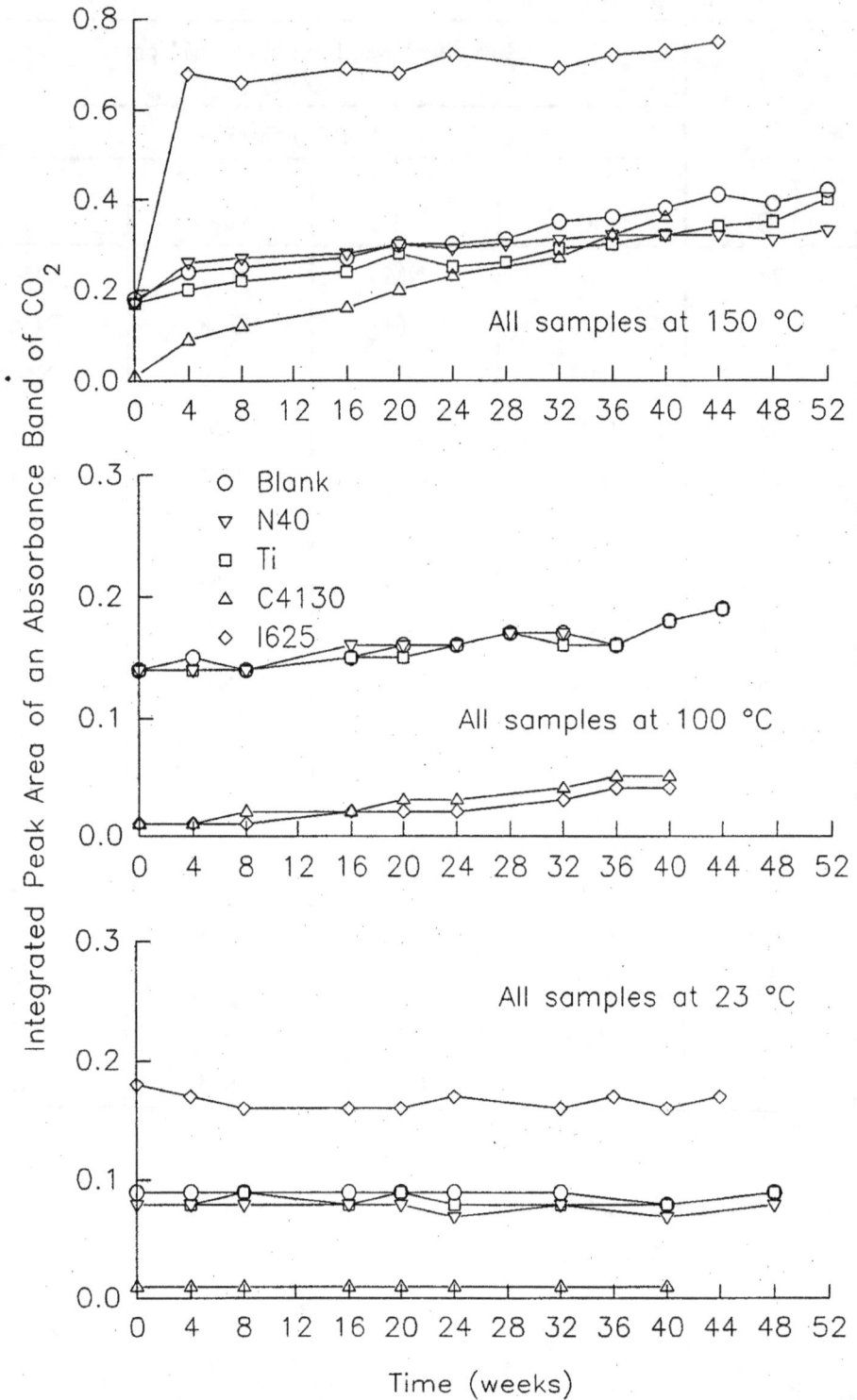

Figure 27. Integrated peak areas for CO_2 for all samples of CF_3I tested in the dry condition without copper at 23 °C, 100 °C, and 150 °C plotted as a function of time.

7. AGENT STABILITY UNDER STORAGE

Table 23. Integrated peak areas of an absorbance band of CO_2 in the 5330 Pa spectra of CF_3I tested in the dry condition with copper

		Integrated area from 667 to 671 cm^{-1} (±0.0048)								
		Week number								
Metal	Temp. (°C)	0	4	8	16	20	24	28	32	36
CDA 110 Blank	23	0.09	0.08	0.08	0.08	0.09	0.08	NM[a]	0.08	NM
CDA 110 N40	23	0.08	0.08	0.08	0.08	0.08	0.08	NM	0.08	NM
CDA 110 Ti	23	0.08	0.08	0.08	0.08	0.08	0.07	NM	0.08	NM
CDA 110 C4130	23	0.01	0.01	0.01	0.01	0.01	0.01	NM	0.01	NM
CDA 110 I625	23	0.17	0.16	0.15	0.15	0.15	0.16	NM	0.15	0.16
CDA 110 Blank	100	0.14	0.14	0.14	0.15	0.15	0.16	0.17	0.18	--[b]
CDA 110 Blank	150	0.17	0.23	0.25	0.27	0.30	0.27	0.28	0.30	0.30
CDA 110 N40	150	0.16	0.23	0.25	0.26	0.29	0.27	0.27	0.29	0.29
CDA 110 Ti	150	0.16	0.22	0.24	0.25	0.28	0.26	0.26	0.27	0.28
CDA 110 C4130	150	0.01	0.12	0.15	0.19	0.20	0.22	NM	0.23	0.26
CDA 110 I625	150	0.17	0.33	0.40	0.48	0.50	0.54	NM	0.54	0.57

Table 23 (cont.). Integrated peak areas of an absorbance band of CO_2 in the 5330 Pa spectra of CF_3I tested in the dry condition with copper

Metal	Temp. (°C)	Integrated area from 667 to 671 cm^{-1} (±0.0048) Week number			
		40	44	48	52
CDA 110 Blank	23	0.08	NM	0.08	NM
CDA 110 N40	23	0.08	NM	0.08	NM
CDA 110 Ti	23	0.08	NM	0.08	NM
CDA 110 C4130	23	0.01	NM	NM	NM
CDA 110 I625	23	0.16	0.16	NM	NM
CDA 110 Blank	100	--[b]	--[b]	--[b]	--[b]
CDA 110 Blank	150	0.30	0.30	0.30	0.32
CDA 110 N40	150	0.30	0.30	0.30	0.32
CDA 110 Ti	150	0.27	0.25	0.27	0.29
CDA 110 C4130	150	0.28	NM	NM	NM
CDA 110 I625	150	0.58	0.59	NM	NM

[a] Not measured
[b] Cylinder emptied after 32 weeks from a leak

7. AGENT STABILITY UNDER STORAGE 311

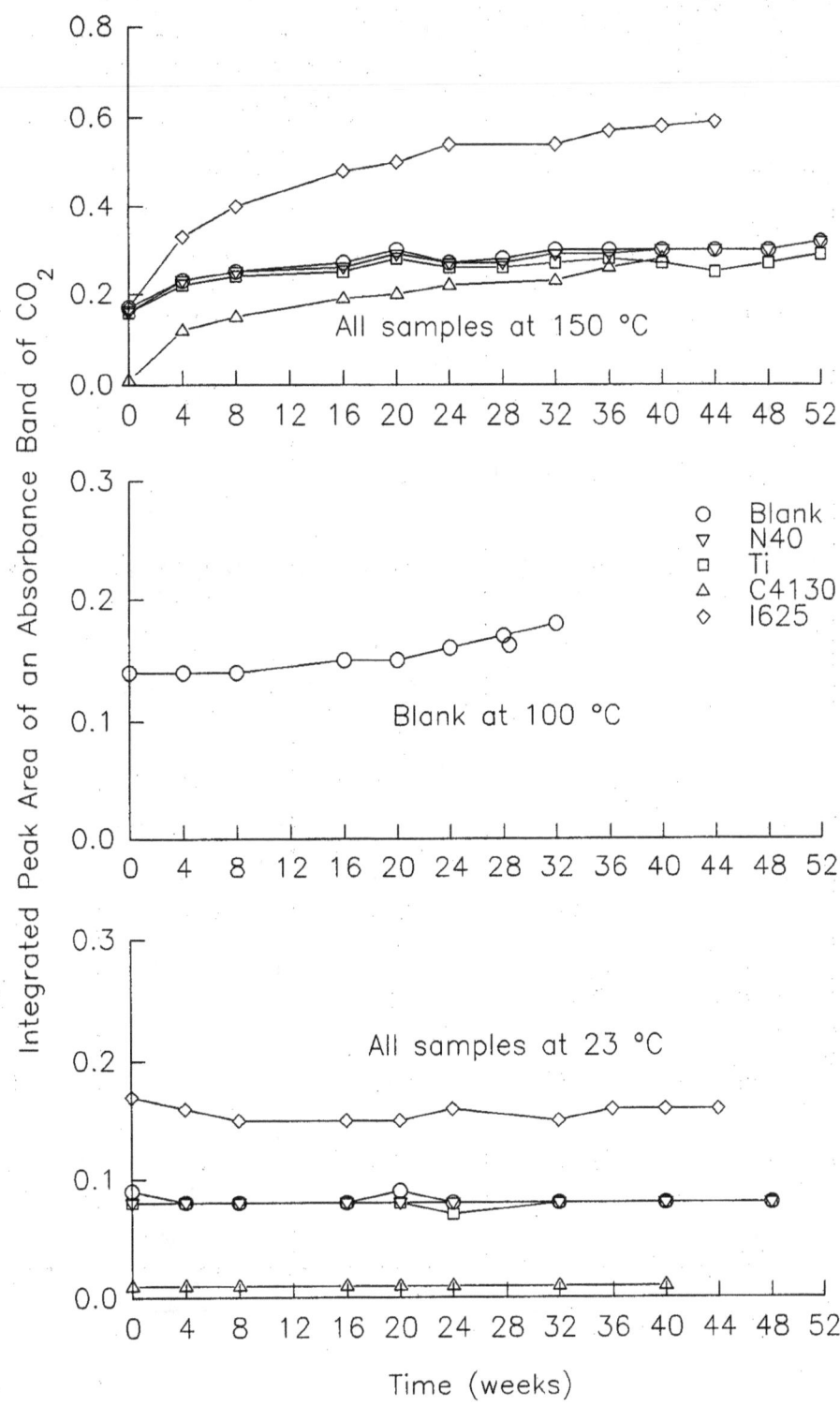

Figure 28. Integrated peak areas for CO_2 for all samples of CF_3I tested in the dry condition with copper at 23 °C, 100 °C, and 150 °C plotted as a function of time.

Table 24. Integrated peak areas for an absorbance band of CO_2 in the 5330 Pa spectra of CF_3I tested in the moist condition without copper

Metal	Temp. (°C)	Integrated area from 667 to 671 cm^{-1} (±0.0048) Week number								
		0	4	8	16	20	24	28	32	36
Blank	100	0.13	0.15	0.16	0.17	0.18	0.18	0.18	0.19	0.18
Blank	150	0.14	0.23	0.26	0.28	0.34	0.37	0.40	0.44	0.43
N40	150	0.14	0.21	0.23	0.23	0.25	0.25	0.25	0.27	0.27
Ti	150	0.13	0.22	0.24	0.25	0.28	0.28	0.29	0.32	0.30
C4130	150	0.01	0.10	0.13	0.20	0.23	0.27	NM[a]	0.29	0.33
I625	150	0.16	0.68	0.66	0.66	0.67	0.70	NM	0.68	0.71

Table 24 (cont.). Integrated peak areas for an absorbance band of CO_2 in the 5330 Pa spectra of CF_3I tested in the moist condition without copper

Metal	Temp. (°C)	Integrated area from 667 to 671 cm^{-1} (±0.0048) Week number			
		40	44	48	52
Blank	100	0.20	0.21	NM	NM
Blank	150	0.45	0.46	0.45	0.49
N40	150	0.27	0.27	0.27	0.29
Ti	150	0.34	0.34	0.34	0.37
C4130	150	0.35	NM	NM	NM
I625	150	0.71	0.75	NM	NM

[a] Not measured.

7. AGENT STABILITY UNDER STORAGE

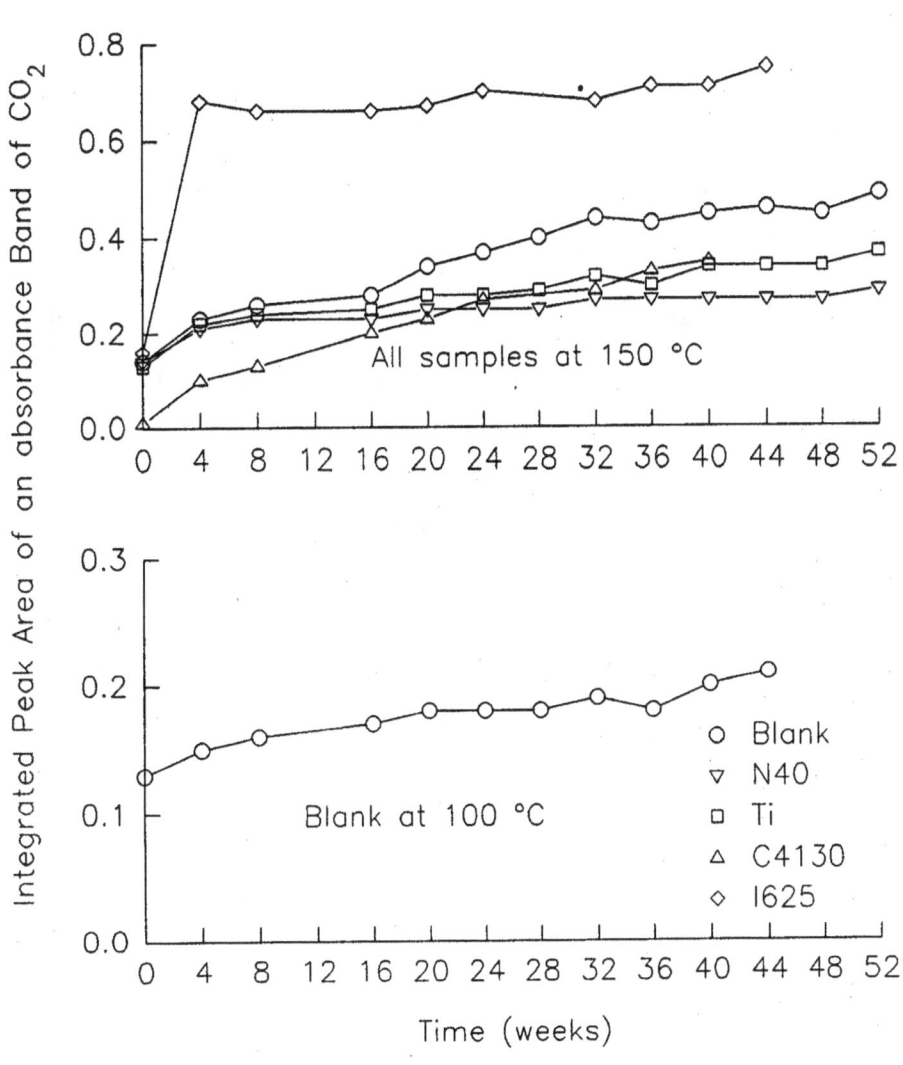

Figure 29. Integrated peak areas for CO_2 for all samples of CF_3I tested in the moist condition without copper at 100 °C and 150 °C plotted as a function of time.

Table 25. Integrated peak areas of an absorbance band of CO_2 in the 5330 Pa spectra of CF_3I tested in the moist condition with copper

Metal	Temp. (°C)	Integrated area from 667 to 671 cm^{-1} (±0.0048) Week number								
		0	4	8	16	20	24	28	32	36
CDA 110 Blank	100	0.13	0.16	0.17	0.18	0.19	0.19	0.19	0.20	0.20
CDA 110 Blank	150	0.13	0.23	0.26	0.28	0.31	0.30	0.32	0.34	0.34
CDA 110 N40	150	0.13	0.22	0.25	0.29	0.33	0.33	0.35	0.37	0.37
CDA 110 Ti	150	0.13	0.20	0.22	0.26	0.29	0.29	0.33	0.36	0.38
CDA 110 C4130	150	0.01	0.13	0.16	0.21	0.22	0.25	NM[a]	0.27	0.30
CDA 110 I625	150	0.01	0.14	0.18	0.25	0.28	0.30	NM	0.32	0.35

Table 25 (cont.). Integrated peak areas of an absorbance band of CO_2 in the 5330 Pa spectra of CF_3I tested in the moist condition with copper

Metal	Temp. (°C)	Integrated area from 667 to 671 cm^{-1} (±0.0048) Week number			
		40	44	48	52
CDA 110 Blank	100	0.21	0.23	NM	NM
CDA 110 Blank	150	0.35	0.34	0.33	0.36
CDA 110 N40	150	0.37	0.38	0.36	0.39
CDA 110 Ti	150	0.39	0.41	0.39	0.43
CDA 110 C4130	150	0.33	NM	NM	NM
CDA 110 I625	150	0.37	NM	NM	NM

[a] Not measured

7. AGENT STABILITY UNDER STORAGE

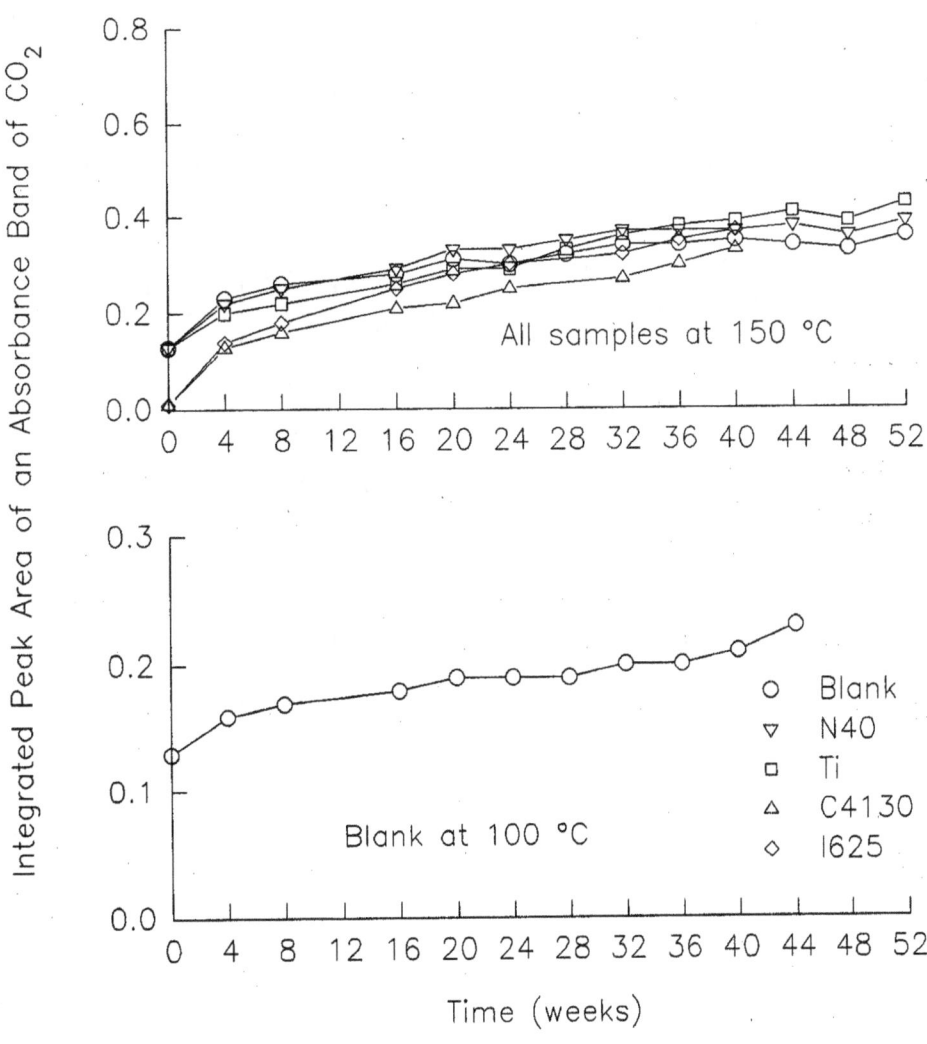

Figure 30. Integrated peak areas for CO_2 for all samples of CF_3I tested in the moist condition with copper at 100 °C and 150 °C plotted as a function of time.

Table 26. Integrated peak areas of an absorbance band for CF_3H in the 5330 Pa spectra of CF_3I tested in the dry condition without copper

Metal	Temp. (°C)	Integrated area from 698 to 701 cm^{-1} (±0.0011)								
		Week number								
		0	4	8	16	20	24	28	32	36
Blank	23	0.01	0.01	0.01	0.01	0.01	0.01	NM^a	0.01	NM
N40	23	0.01	0.01	0.01	0.01	0.01	0.01	NM	0.01	NM
Ti	23	--^{b}	0.01	0.01	0.01	0.01	0.01	NM	0.01	NM
C4130	23	ND^c	ND	ND	ND	ND	ND	NM	ND	NM
I625	23	0.04	0.04	0.04	0.04	0.04	0.04	NM	0.04	0.04
Blank	100	0.03	0.03	0.03	0.03	0.03	0.03	0.03	0.03	0.03
N40	100	0.03	0.03	0.03	0.03	0.03	0.03	0.03	0.03	--^{d}
Ti	100	0.03	0.03	0.03	0.03	0.03	0.03	0.03	0.03	0.03
C4130	100	ND	ND	ND	ND	ND	ND	NM	ND	ND
I625	100	ND	ND	ND	ND	ND	ND	NM	ND	ND
Blank	150	0.02	0.02	0.03	0.04	0.05	0.06	0.08	0.11	0.15
N40	150	0.02	0.03	0.04	0.06	0.07	0.09	0.10	0.11	0.12
Ti	150	0.02	0.02	0.03	0.03	0.03	0.03	0.04	0.04	0.05
C4130	150	ND	<0.01	0.01	0.03	0.04	0.06	NM	0.12	0.15
I625	150	0.04	0.06	0.10	0.22	0.26	0.30	NM	0.35	0.37

7. AGENT STABILITY UNDER STORAGE

Table 26 (cont.). Integrated peak areas of an absorbance band for CF_3H in the 5330 Pa spectra of CF_3I tested in the dry condition without copper

Metal	Temp. (°C)	Integrated area from 698 to 701 cm^{-1} (±0.0011) Week number			
		40	44	48	52
Blank	23	0.01	NM	0.01	NM
N40	23	0.01	NM	0.01	NM
Ti	23	0.01	NM	0.01	NM
C4130	23	ND	NM	NM	NM
I625	23	0.04	0.04	NM	NM
Blank	100	0.03	0.04	NM	NM
N40	100	--^{d}	--^{d}	--^{d}	--^{d}
Ti	100	0.03	0.04	NM	NM
C4130	100	ND	NM	NM	NM
I625	100	ND	NM	NM	NM
Blank	150	0.19	0.24	0.28	0.32
N40	150	0.12	0.13	0.14	0.15
Ti	150	0.06	0.07	0.10	0.12
C4130	150	0.19	NM	NM	NM
I625	150	0.39	0.41	NM	NM

[a] Not measured
[b] Lost spectrum file
[c] Not detected
[d] Cylinder emptied after 32 weeks from a leak.

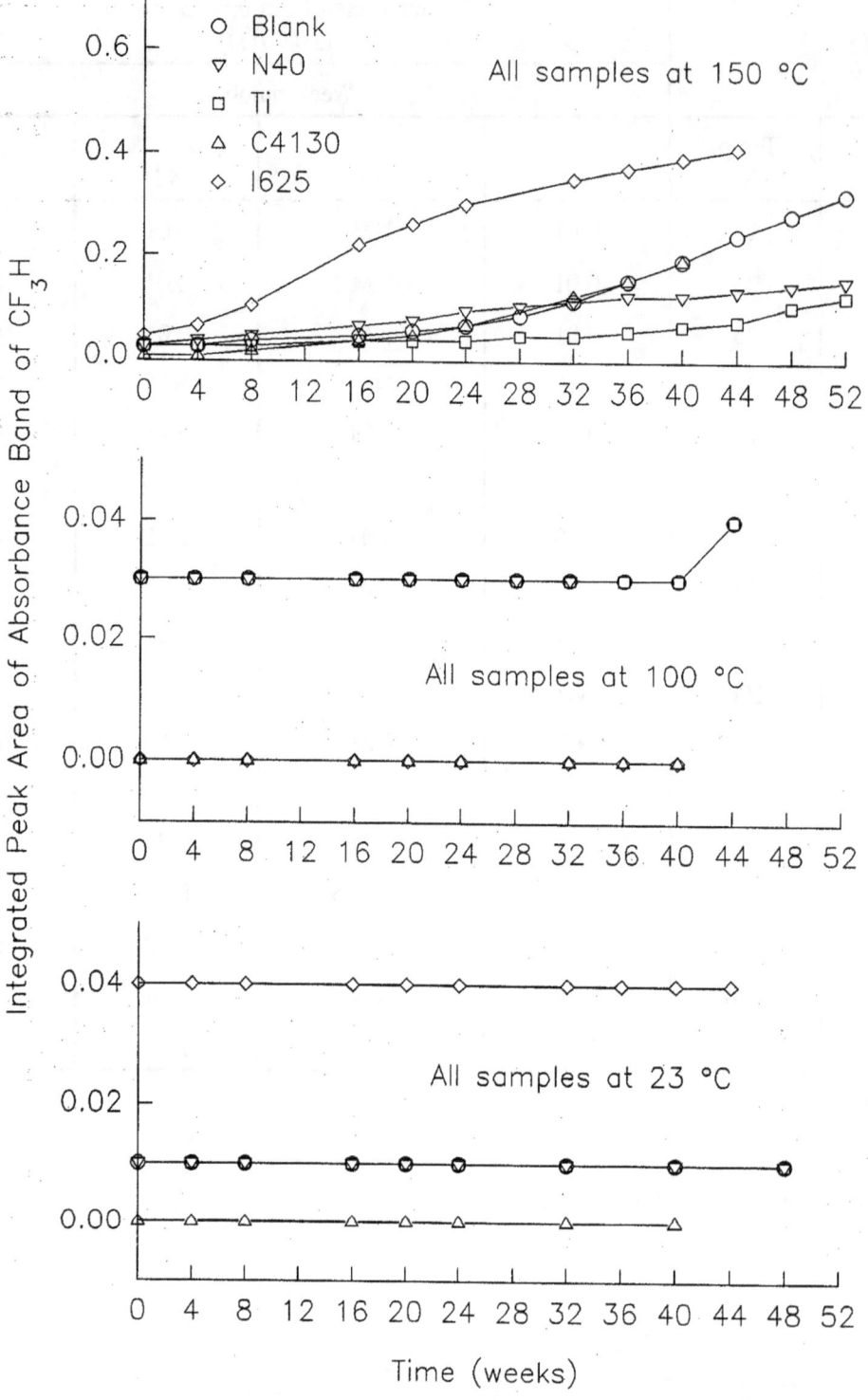

Figure 31. Integrated peak areas for CF_3H for all samples of CF_3I tested in the dry condition without copper at 23 °C, 100 °C, and 150 °C plotted as a function of time.

7. AGENT STABILITY UNDER STORAGE

Table 27. Integrated peak areas of an absorbance band for CF_3H in the 5330 Pa spectra of CF_3I tested in the dry condition with copper

Metal	Temp. (°C)	Integrated area from 698 to 701 cm^{-1} (±0.0011)								
		Week number								
		0	4	8	16	20	24	28	32	36
CDA 110 Blank	23	0.01	0.01	0.01	0.01	0.01	0.01	NM	0.01	NM
CDA 110 N40	23	0.01	0.01	0.01	0.01	0.01	0.01	NM[a]	0.01	NM
CDA 110 Ti	23	0.01	0.01	0.01	0.01	0.01	0.01	NM	0.01	NM
CDA 110 C4130	23	ND[b]	ND	ND	ND	ND	ND	NM	ND	NM
CDA 110 I625	23	0.04	0.04	0.04	0.04	0.04	0.04	NM	0.04	0.04
CDA 110 Blank	100	0.03	0.03	0.03	0.03	0.03	0.03	0.03	0.03	--[c]
CDA 110 Blank	150	0.02	0.02	0.03	0.06	0.07	0.09	0.12	0.15	0.19
CDA 110 N40	150	0.02	0.03	0.05	0.16	0.22	0.27	0.32	0.37	0.39
CDA 110 Ti	150	0.02	0.02	0.03	0.05	0.07	0.09	0.10	0.12	0.14
CDA 110 C4130	150	ND	0.01	0.02	0.03	0.05	0.05	NM	0.07	0.09
CDA 110 I625	150	0.04	0.04	0.04	0.05	0.07	0.08	NM	0.13	0.16

Table 27 (cont.). Integrated peak areas of an absorbance band for CF_3H in the 5330 Pa spectra of CF_3I tested in the dry condition with copper

Metal	Temp. (°C)	Integrated area from 698 to 701 cm^{-1} (±0.0011)			
		Week number			
		40	44	48	52
CDA 110 Blank	23	0.01	NM	0.01	NM
CDA 110 N40	23	0.01	NM	0.01	NM
CDA 110 Ti	23	0.01	NM	0.01	NM
CDA 110 C4130	23	ND	NM	NM	NM
CDA 110 I625	23	0.04	0.04	NM	NM
CDA 110 Blank	100	--^{c}	--^{c}	--^{c}	--^{c}
CDA 110 Blank	150	0.23	0.28	0.33	0.37
CDA 110 N40	150	0.41	0.45	0.47	0.49
CDA 110 Ti	150	0.16	0.18	0.20	0.22
CDA 110 C4130	150	0.11	NM	NM	NM
CDA 110 I625	150	0.21	0.23	NM	NM

^{a} Cylinder emptied after 32 weeks from a leak

7. AGENT STABILITY UNDER STORAGE

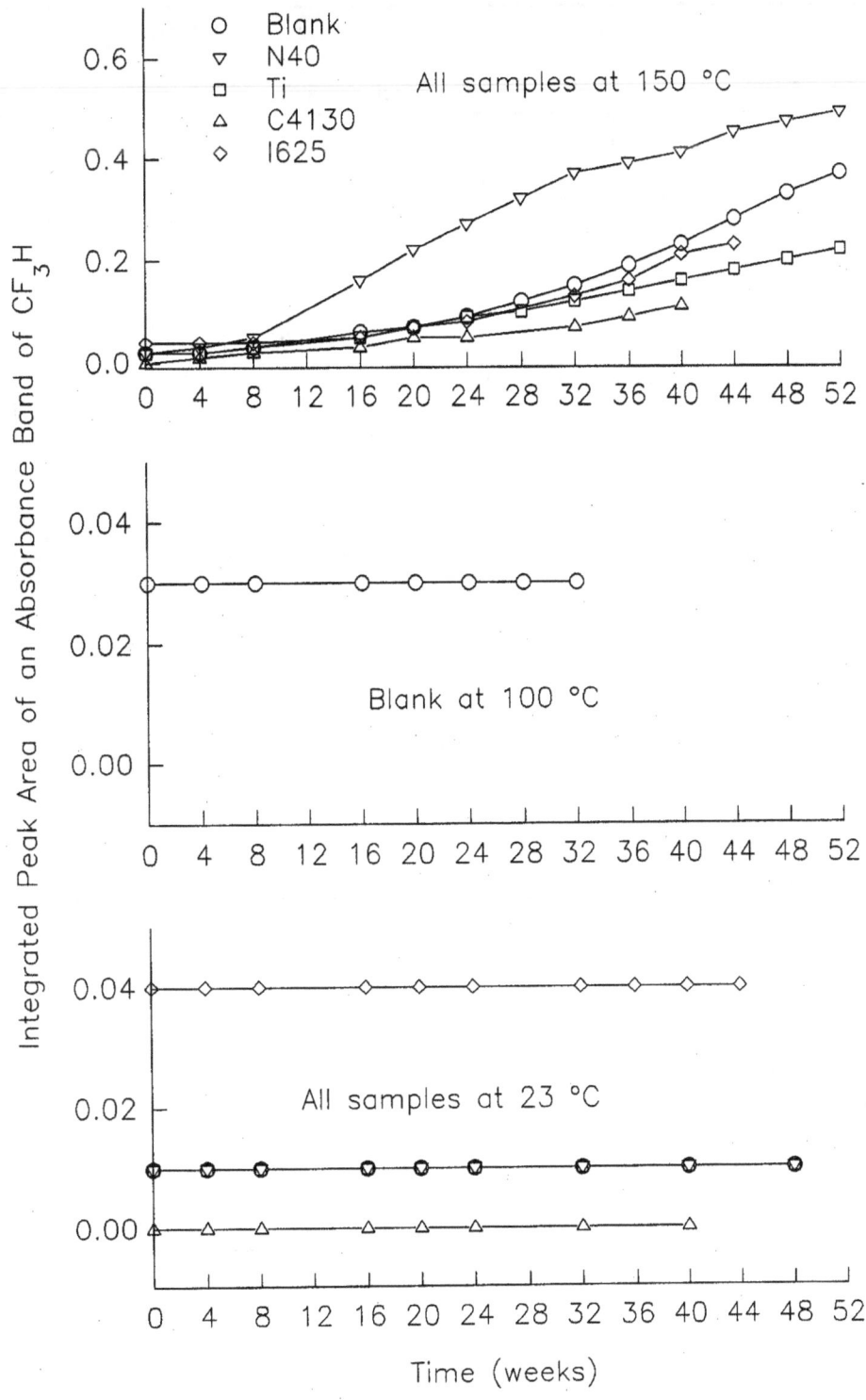

Figure 32. Integrated peak areas for CF_3H for all samples of CF_3I tested in the dry condition with copper at 23 °C, 100 °C, and 150 °C plotted as a function of time.

Table 28. Integrated peak areas of an absorbance band for CF_3H in the 5330 Pa spectra of CF_3I tested in the moist condition without copper

Metal	Temp. (°C)	Integrated area from 698 to 701 cm^{-1} (±0.0011)								
		Week number								
		0	4	8	16	20	24	28	32	36
Blank	100	0.03	0.03	0.03	0.03	0.03	0.03	0.03	0.03	0.03
Blank	150	0.01	0.02	0.03	0.07	0.11	0.18	0.24	0.29	0.34
N40	150	0.01	0.03	0.05	0.08	0.10	0.12	0.14	0.15	0.16
Ti	150	0.02	0.02	0.03	0.05	0.07	0.09	0.12	0.14	0.17
C4130	150	ND[a]	0.01	0.01	0.05	0.08	0.11	NM[b]	0.18	0.22
I625	150	0.04	0.12	0.25	0.36	0.38	0.41	NM	0.44	0.46

Table 28 (cont.). Integrated peak areas of an absorbance band for CF_3H in the 5330 Pa spectra of CF_3I tested in the moist condition without copper

Metal	Temp. (°C)	Integrated area from 698 to 701 cm^{-1} (±0.0011)			
		Week number			
		40	44	48	52
Blank	100	0.03	0.03	NM	NM
Blank	150	0.37	0.40	0.42	0.45
N40	150	0.17	0.19	0.20	0.21
Ti	150	0.19	0.22	0.24	0.27
C4130	150	0.27	NM	NM	NM
I625	150	0.48	0.51	NM	NM

[a] Not detected
[b] Not measured.

7. AGENT STABILITY UNDER STORAGE

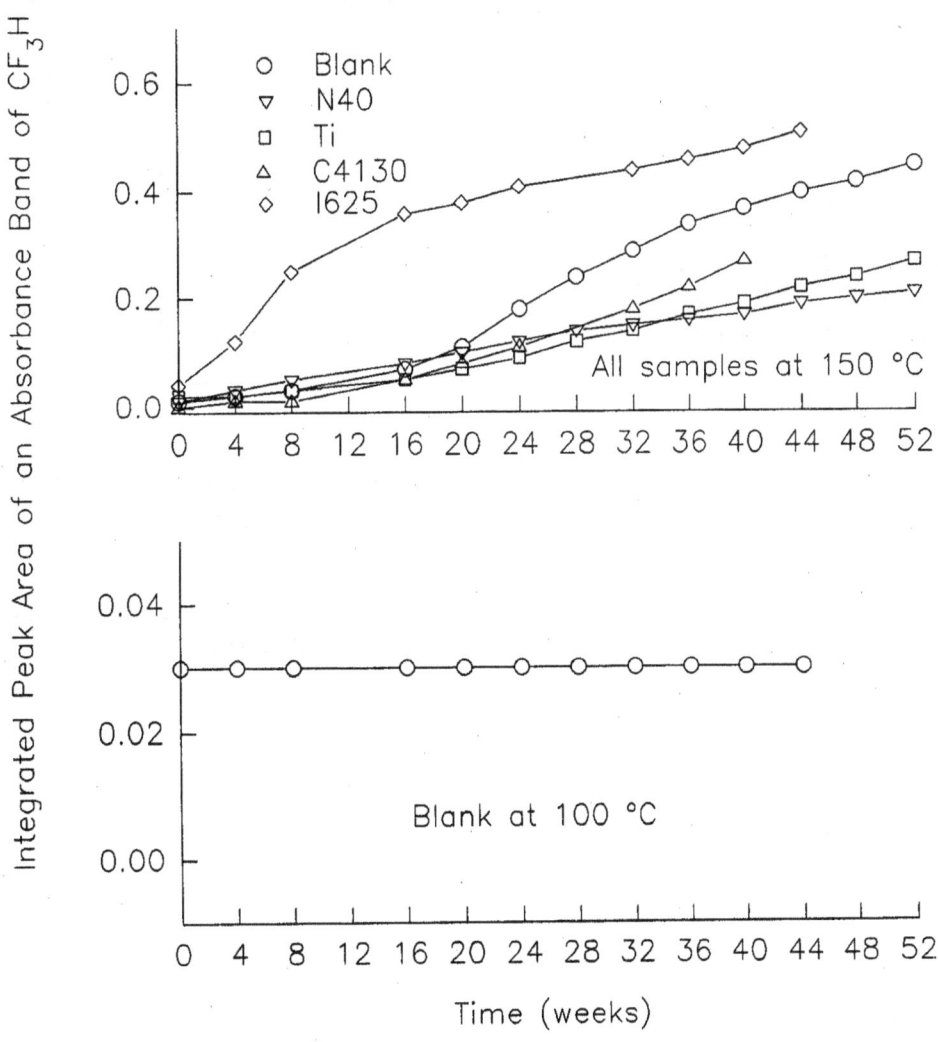

Figure 33. Integrated peak areas for CF_3H for all samples of CF_3I tested in the moist condition without copper at 100 °C and 150 °C plotted as a function of time.

Table 29. Integrated peak areas of an absorbance band for CF_3H in the 5330 Pa spectra of CF_3I tested in the moist condition with copper

Metal	Temp. (°C)	Integrated area from 698 to 701 cm^{-1} (±0.0011) Week number								
		0	4	8	16	20	24	28	32	36
CDA 110 Blank	100	0.03	0.03	0.03	0.03	0.03	0.03	0.03	0.03	0.03
CDA 110 Blank	150	0.01	0.02	0.02	0.06	0.09	0.14	0.19	0.24	0.29
CDA 110 N40	150	0.01	0.02	0.03	0.07	0.10	0.14	0.19	0.26	0.35
CDA 110 Ti	150	0.01	0.02	0.05	0.09	0.11	0.13	0.15	0.16	0.18
CDA 110 C4130	150	<0.01	0.01	0.02	0.04	0.05	0.05	NM[a]	0.06	0.07
CDA 110 I625	150	0.03	0.01	0.01	0.03	0.05	0.07	NM	0.13	0.16

Table 29 (cont.). Integrated Peak Areas of an Absorbance Band for CF_3H in the 5330 Pa Spectra of CF_3I tested in the Moist Condition with Copper

Metal	Temp. (°C)	Integrated Area from 698 to 701 cm^{-1} (±0.0011) Week Number			
		40	44	48	52
CDA 110 Blank	100	0.03	0.03	NM	NM
CDA 110 Blank	150	0.34	0.37	0.40	0.44
CDA 110 N40	150	0.41	0.50	0.55	0.63
CDA 110 Ti	150	0.19	0.20	0.22	0.24
CDA 110 C4130	150	0.08	NM	NM	NM
CDA 110 I625	150	0.21	NM	NM	NM

[a] Not measured

7. AGENT STABILITY UNDER STORAGE

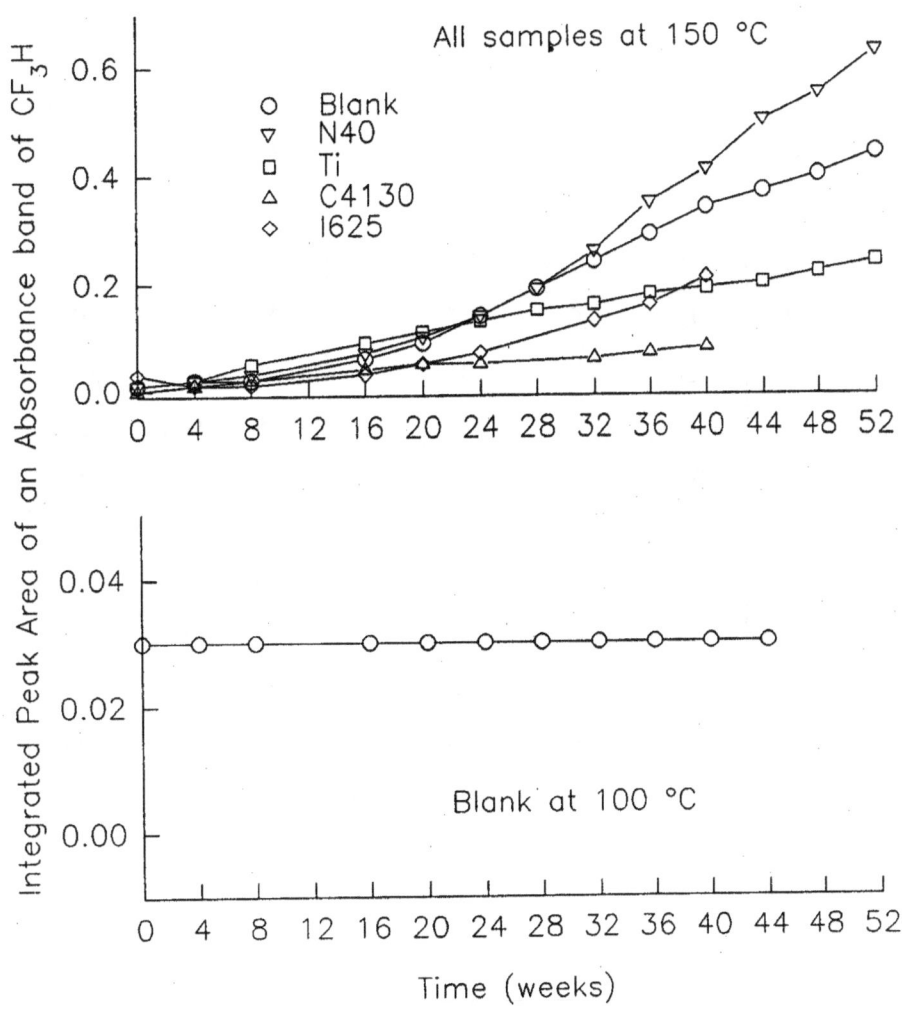

Figure 34. Integrated peak areas for CF_3H for all samples of CF_3I tested in the moist condition with copper at 100 °C and 150 °C plotted as a function of time.

Table 30. Integrated peak areas for an absorbance band of the fluorinated alkene in the 5330 Pa spectra of CF_3I tested in the dry condition without copper

Metal	Temp. (°C)	Integrated area from 944 to 953 cm^{-1} (±0.011)								
		Week number								
		0	4	8	16	20	24	28	32	36
Blank	23	ND[a]	ND	ND	ND	ND	ND	NM[b]	ND	NM
N40	23	ND	ND	ND	ND	ND	ND	NM	ND	NM
Ti	23	--[c]	ND	ND	ND	ND	ND	NM	ND	NM
C4130	23	ND	ND	ND	ND	ND	ND	NM	ND	NM
I625	23	ND	ND	ND	ND	ND	ND	NM	ND	ND
Blank	100	ND	ND	ND	0.01	0.01	0.01	.02	.02	0.02
N40	100	ND	<0.01	0.01	0.02	0.03	0.03	.03	.04	--[d]
Ti	100	ND	ND	0.01	0.02	0.02	0.03	.04	.04	0.05
C4130	100	ND	<0.01	0.01	0.04	0.05	0.06	NM	0.06	0.07
I625	100	ND	ND	0.01	0.02	0.03	0.03	NM	0.06	0.07
Blank	150	ND	0.06	0.09	0.10	0.10	0.11	0.10	0.11	0.11
N40	150	ND	0.01	0.10	0.10	0.11	0.11	0.11	0.11	0.11
Ti	150	ND	0.03	0.06	0.07	0.12	0.08	0.08	0.09	0.09
C4130	150	ND	0.06	0.09	0.11	0.11	0.12	NM	0.11	0.12
I625	150	ND	0.06	0.06	0.12	0.17	0.12	NM	0.15	0.13

Table 30 (cont.). Integrated peak areas for an absorbance band of the fluorinated alkene in the 5330 Pa spectra of CF_3I tested in the dry condition without copper

		Integrated area from 944 to 953 cm^{-1} (±0.011)			
		Week number			
Metal	Temp. (°C)	40	44	48	52
Blank	23	ND	NM	ND	NM
N40	23	ND	NM	ND	NM
Ti	23	ND	NM	ND	NM
C4130	23	ND	NM	NM	NM
I625	23	ND	ND	NM	NM
Blank	100	0.02	0.03	NM	NM
N40	100	--^{d}	--^{d}	--^{d}	--^{d}
Ti	100	0.05	0.06	NM	NM
C4130	100	0.07	NM	NM	NM
I625	100	0.08	NM	NM	NM
Blank	150	0.11	0.11	0.11	0.11
N40	150	0.12	0.12	0.12	0.12
Ti	150	0.10	0.10	0.10	0.10
C4130	150	0.14	NM	NM	NM
I625	150	0.12	0.12	NM	NM

[a] Not detected
[b] Not measured
[c] Lost spectrum file
[d] Cylinder emptied after 32 weeks from a leak.

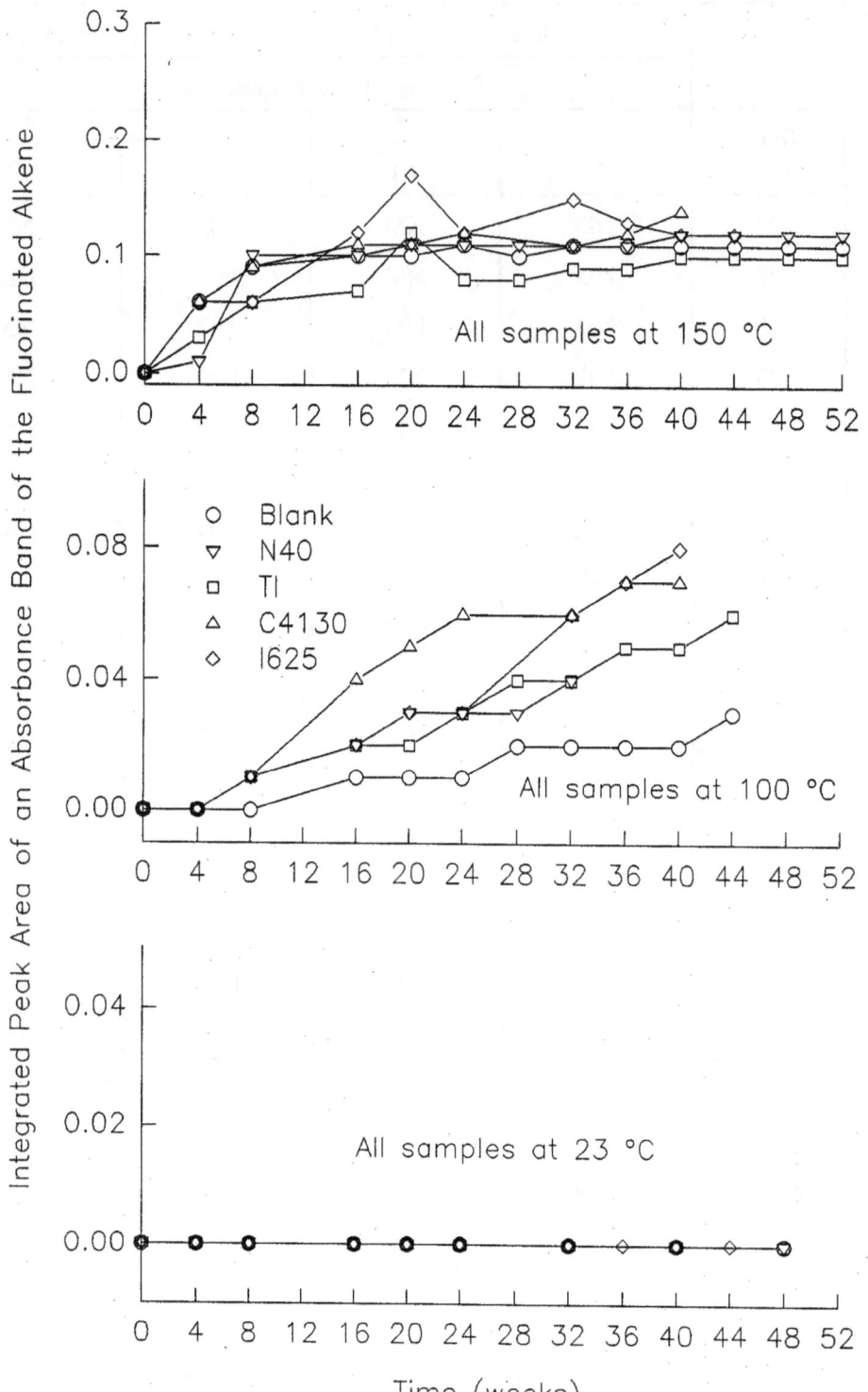

Figure 35. Integrated peak areas for the fluorinated alkene for all samples of CF_3I tested in the dry condition without copper at 23 °C, 100 °C, and 150 °C plotted as a function of time.

Table 31. Integrated peak areas for an absorbance band of the fluorinated alkene in the 5330 Pa spectra of CF_3I tested in the dry condition with copper

		Integrated area from 944 to 953 cm^{-1} (±0.011)								
		Week number								
Metal	Temp. (°C)	0	4	8	16	20	24	28	32	36
CDA 110 Blank	23	ND[a]	ND	ND	ND	ND	ND	NM[b]	ND	NM
CDA 110 N40	23	ND	ND	ND	ND	ND	ND	NM	ND	NM
CDA 110 Ti	23	ND	ND	ND	ND	ND	ND	NM	ND	NM
CDA 110 C4130	23	ND	ND	ND	ND	ND	ND	NM	ND	NM
CDA 110 I625	23	ND	ND	ND	ND	ND	ND	NM	ND	ND
CDA 110 Blank	100	ND	ND	ND	0.01	0.01	0.01	.02	.02	--[c]
CDA 110 Blank	150	ND	0.11	0.14	0.08	0.08	0.06	0.06	0.06	0.05
CDA 110 N40	150	ND	0.13	0.14	0.13	0.14	0.12	0.12	0.12	0.10
CDA 110 Ti	150	ND	0.10	0.12	0.10	0.09	0.07	0.06	0.05	0.04
CDA 110 C4130	150	ND	0.11	0.14	0.18	0.19	0.19	NM	0.19	0.19
CDA 110 I625	150	ND	0.04	0.10	0.14	0.16	0.14	NM	0.14	0.12

Table 31 (cont.). Integrated peak areas for an absorbance band of the fluorinated alkene in the 5330 Pa spectra of CF_3I tested in the dry condition with copper

Metal	Temp. (°C)	Integrated area from 944 to 953 cm^{-1} (±0.011) Week number			
		40	44	48	52
CDA 110 Blank	23	ND	NM	ND	NM
CDA 110 N40	23	ND	NM	ND	NM
CDA 110 Ti	23	ND	NM	ND	NM
CDA 110 C4130	23	ND	NM	NM	NM
CDA 110 I625	23	ND	ND	NM	NM
CDA 110 Blank	100	--^{c}	--^{c}	--^{c}	--^{c}
CDA 110 Blank	150	0.05	0.04	0.04	0.03
CDA 110 N40	150	0.10	0.10	0.09	0.08
CDA 110 Ti	150	0.03	0.03	0.02	0.02
CDA 110 C4130	150	0.19	NM	NM	NM
CDA 110 I625	150	0.12	0.12	NM	NM

[a] Not detected
[b] Not measured
[c] Cylinder emptied after 32 weeks from a leak.

7. AGENT STABILITY UNDER STORAGE

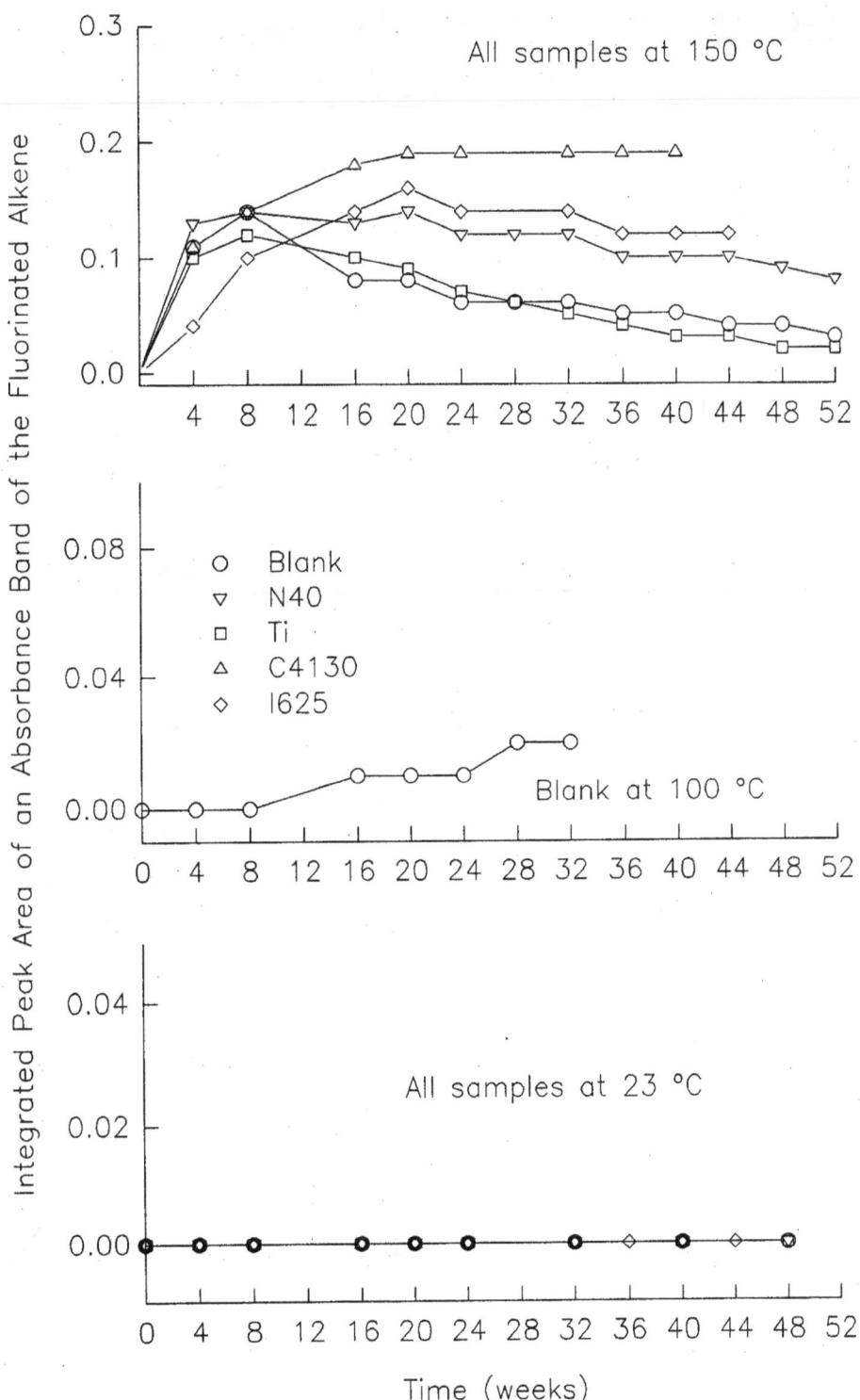

Figure 36. Integrated peak areas for the fluorinated alkene for all samples of CF$_3$I tested in the dry condition with copper at 23 °C, 100 °C, and 150 °C plotted as a function of time.

Table 32. Integrated peak areas of an absorbance band of the fluorinated alkene in the 5330 Pa spectra of CF_3I tested in the moist condition without copper

Metal	Temp. (°C)	Integrated area from 944 to 953 cm^{-1} (±0.011)								
		Week number								
		0	4	8	16	20	24	28	32	36
Blank	100	ND[a]	ND	0.01	0.02	0.03	0.03	0.04	0.04	0.04
Blank	150	ND	0.08	0.10	0.16	0.20	0.18	0.17	0.18	0.18
N40	150	ND	0.09	0.10	0.10	0.11	0.12	0.11	0.12	0.12
Ti	150	ND	0.08	0.07	0.10	0.10	0.10	0.09	0.09	0.09
C4130	150	ND	0.11	0.13	0.23	0.24	0.19	NM[b]	0.19	0.18
I625	150	ND	0.12	0.14	0.12	0.13	0.12	NM	0.12	0.12

Table 32 (cont.). Integrated peak areas of an absorbance band of the fluorinated alkene in the 5330 Pa spectra of CF_3I tested in the moist condition without copper

Metal	Temp. (°C)	Integrated area from 944 to 953 cm^{-1} (±0.011)			
		Week number			
		40	44	48	52
Blank	100	0.04	0.05	NM	NM
Blank	150	0.17	0.18	0.18	0.19
N40	150	0.12	0.12	0.12	0.12
Ti	150	0.09	0.08	0.09	0.09
C4130	150	0.20	NM	NM	NM
I625	150	0.12	0.13	NM	NM

[a] Not detected
[b] Not measured.

7. AGENT STABILITY UNDER STORAGE

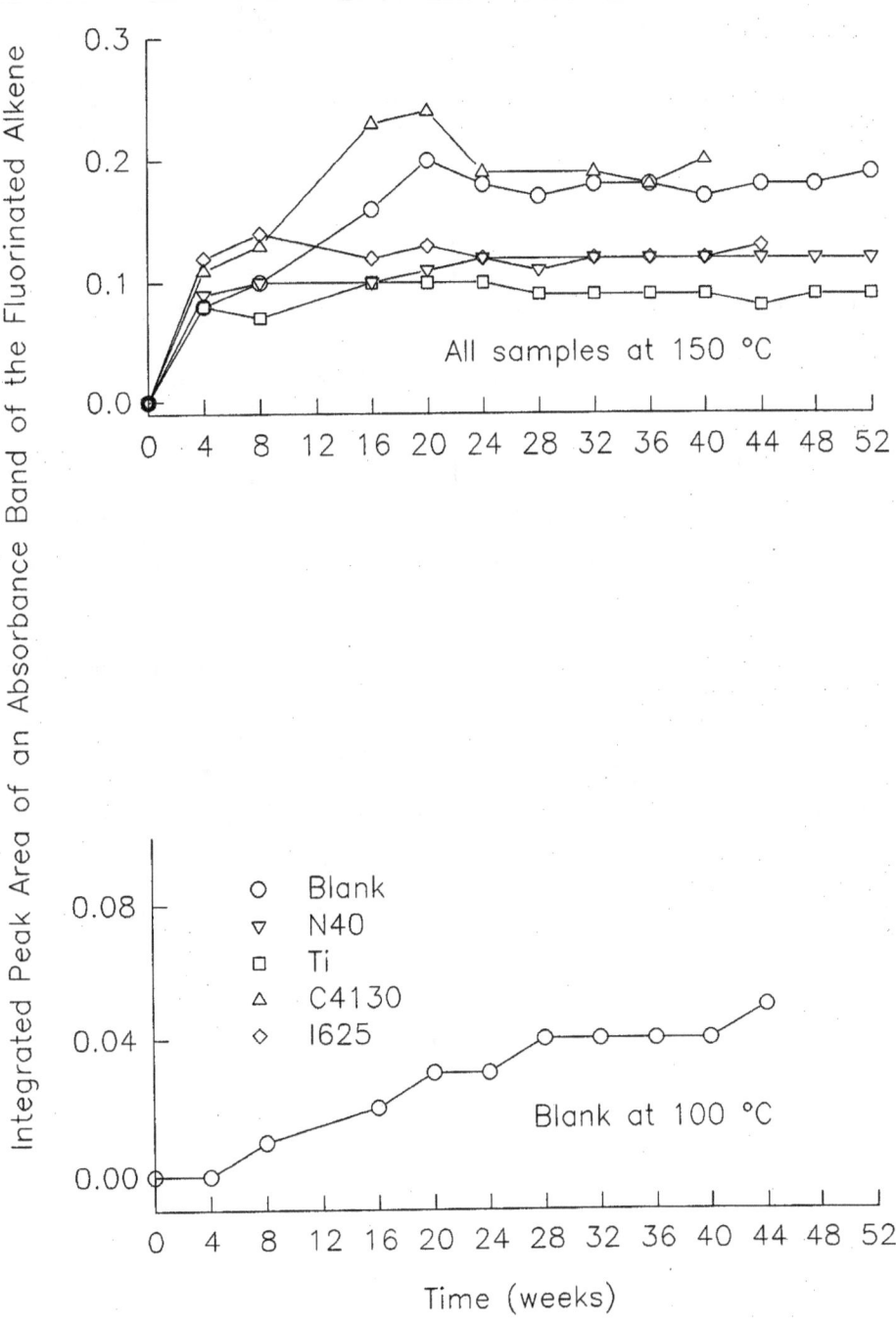

Figure 37. Integrated peak areas for the fluorinated alkene for all samples of CF_3I tested in the moist condition without copper at 100 °C and 150 °C plotted as a function of time.

Table 33. Integrated peak areas of an absorbance band of the fluorinated alkene in the 5330 Pa spectra of CF_3I tested in the moist condition with copper

Metal	Temp. (°C)	Integrated area from 944 to 953 cm^{-1} (±0.011) Week number								
		0	4	8	16	20	24	28	32	36
CDA 110 Blank	100	ND[a]	ND	0.01	0.01	0.02	0.02	0.02	0.02	0.03
CDA 110 Blank	150	ND	ND	ND	ND	ND	ND	ND	ND	ND
CDA 110 N40	150	ND	0.02	0.04	0.11	0.12	0.12	0.12	0.12	0.12
CDA 110 Ti	150	ND	0.04	0.07	0.04	0.03	0.03	0.03	0.03	0.03
CDA 110 C4130	150	ND	0.12	0.18	0.22	0.23	0.17	NM[b]	0.18	0.17
CDA 110 I625	150	ND	0.11	0.04	0.06	0.04	0.03	NM	0.03	0.03

Table 33 (cont.). Integrated peak areas of an absorbance band of the fluorinated alkene in the 5330 Pa spectra of CF_3I tested in the moist condition with copper

Metal	Temp. (°C)	Integrated area from 944 to 953 cm^{-1} (±0.011) Week number			
		40	44	48	52
CDA 110 Blank	100	0.03	0.03	NM	NM
CDA 110 Blank	150	ND	ND	ND	ND
CDA 110 N40	150	0.12	0.12	0.12	0.12
CDA 110 Ti	150	0.03	0.03	0.03	0.03
CDA 110 C4130	150	0.18	NM	NM	NM
CDA 110 I625	150	0.04	NM	NM	NM

[a] Not detected
[b] Not measured.

7. AGENT STABILITY UNDER STORAGE

Table 34. Summary of the initial and final peak areas of CF$_3$I at all temperatures and conditions

Metal	Number of weeks	Temp. (°C)	Area (± 0.011) Init.	Area (± 0.011) Final	Significant @ 3 x uncertainty
Condition: Dry without Copper					
blank	48	23	0.25	0.23	no
N40	48	23	0.19[a]	0.24	--
Ti	48	23	0.26	0.23	no
C4130	40	23	0.24	0.25	no
I625	44	23	0.27	0.24	no
blank	44	100	0.25	0.25	no
N40	32	100	0.24	0.24	no
Ti	44	100	0.25	0.24	no
C4130	40	100	0.25	0.24	no
I625	40	100	0.25	0.24	no
blank	52	150	0.28	0.25	no
N40	52	150	0.27	0.25	no
Ti	52	150	0.28	0.25	no
C4130	40	150	0.29[a]	0.23	--
I625	44	150	0.25	0.22	no
Condition: Dry with Copper					
blank	48	23	0.26	0.23	no
N40	48	23	0.26	0.24	no
Ti	48	23	0.27	0.25	no
C4130	40	23	0.23	0.24	no
I625	44	23	0.26	0.23	no
blank	32	100	0.25	0.26	no
blank	52	150	0.29	0.23	yes
N40	52	150	0.26	0.19	yes
Ti	52	150	0.28	0.23	yes
C4130	40	150	0.24	0.22	no
I625	44	150	0.25	0.23	no

Table 34. (continued) Summary of the initial and final peak areas of CF_3I at all temperatures and conditions

Metal	Number of weeks	Temp. °C	Area (± 0.011) Init.	Area (± 0.011) Final	Significant @ 3 x uncertainty
Condition: Moist without copper					
blank	44	100	0.25	0.24	no
blank	52	150	0.27	0.23	yes
N40	52	150	0.28	0.23	yes
Ti	52	150	0.28	0.23	yes
C4130	40	150	0.24	0.21	no
I625	44	150	0.25	0.21	yes
Condition: Moist with copper					
blank	44	100	0.25	0.24	no
blank	52	150	0.27	0.22	yes
N40	52	150	0.27	0.19	yes
Ti	52	150	0.28	0.21	yes
C4130	40	150	0.23	0.21	no
I625	40	150	0.23	0.21	no

[a] Suspected to be an outlier because of improper gas cell filling.

Table 35. Spectral comparisons of initial and aged CF_3I samples

Condition		Without copper (± 0.0018)			With copper (± 0.0018)		
Metal	Weeks tested	Dry at 23 °C	Dry at 150 °C	Moist at 150 °C	Dry at 23 °C	Dry at 150 °C	Moist at 150 °C
Blank	52	0.996	0.991	0.984	0.997	0.980	0.981
N40	52	0.998	0.995	0.992	0.997	0.875	0.949
Ti	52	0.998	0.996	0.990	0.997	0.985	0.990
C4130	40	0.999	0.994	0.990	0.999	0.997	0.998
I625	44	0.998	0.988	0.983	0.999	0.994	0.994

7. AGENT STABILITY UNDER STORAGE

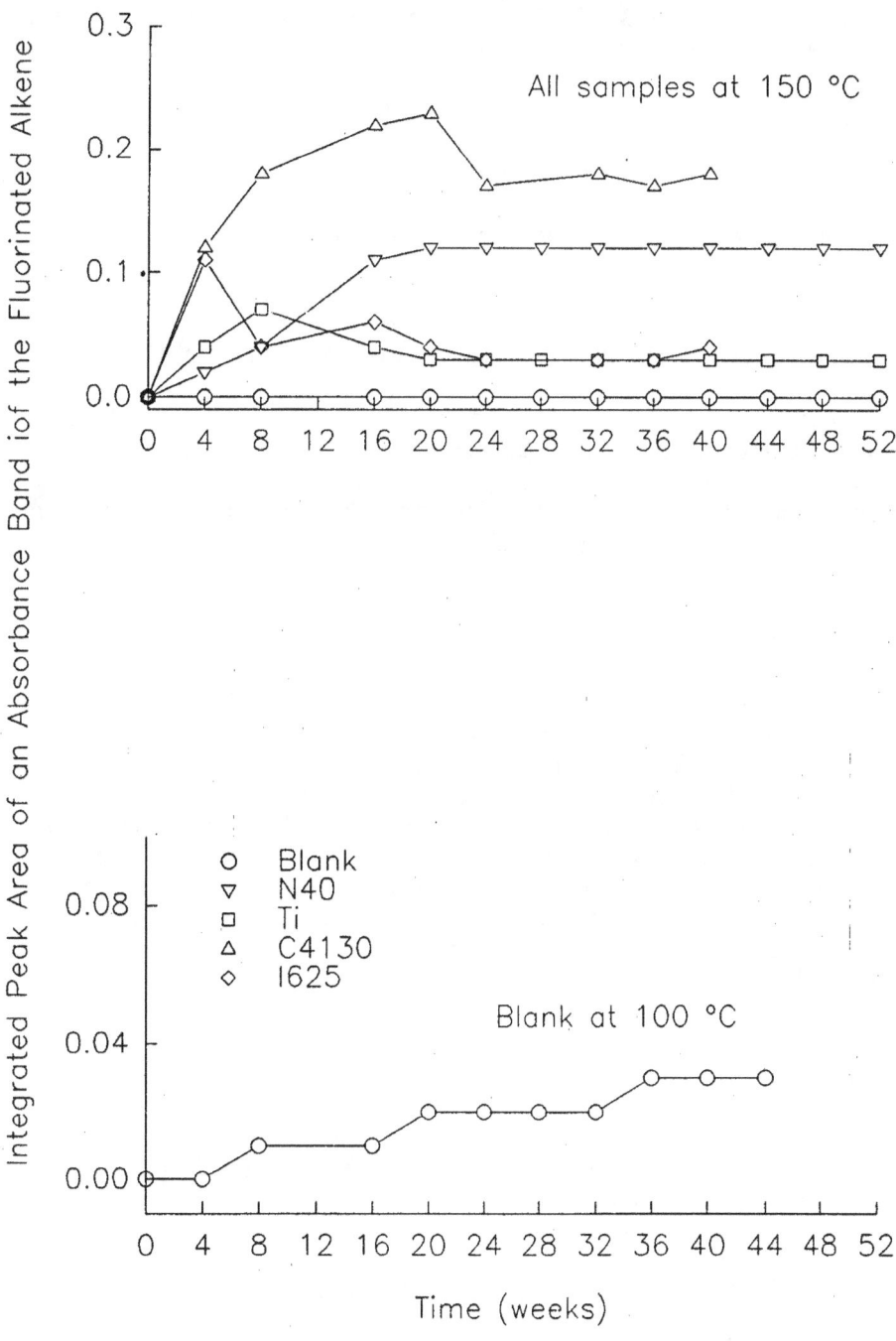

Figure 38. Integrated peak areas for the fluorinated alkene for all samples of CF_3I tested in the moist condition with copper at 100 °C and 150 °C plotted as a function of time.

7.4 Conclusions

The long-term stability project showed the following:

- the fluorocarbon agents FC-218, HFC-125, and HFC-227ea were stable at temperatures as high as 150 °C for as long as 48 weeks. No by-products were formed.

- CF_3I degraded at 100 °C and was accelerated at 150 °C.

- CF_3H, CO_2, and CO were produced in low levels as degradation products of CF_3I.

- the presence of moisture accelerates the degradation of CF_3I.

- the presence of copper accelerates the degradation of CF_3I.

- the presence of copper and moisture accelerate the degradation of CF_3I.

- an absorbance band at 950 cm^{-1} was generated in the CF_3I samples that may be from a fluorinated alkene; the presence of copper at 150 °C caused the double bond to break.

- storage at ambient conditions of any of the four agents is feasible, but storage at elevated temperatures for CF_3I needs more study.

7.5 Acknowledgments

The author wishes to acknowledge a number of additional NIST staff members who made contributions to this section. These include Mr. Richard Peacock, Dr. Marc Nyden, Mr. Thomas Cleary, Dr. Takashi Kashiwagi, Dr. Richard Gann, and Mr. Darren Lowe for providing their technical expertise. Also, Ms. Paula Garrett for organizing and editing the section.

In addition the author wishes to acknowledge Mr. Richard Sears of Walter Kidde Aerospace for coordinating the treatment of the C4130 alloy steel and Mr. Larry Braden of Metal Samples Co., Inc. for assistance in providing the numerous metal coupons used. Dr. Douglas Dierdorf of Pacific Scientific helped to try and identify and confirm the presence of a fluorinated alkene in aged CF_3I samples. Mr. Thomas Austin of Washington Valve and Fitting Co. recommended equipment that made the FTIR analyses procedure very precise.

7.6 References

Balzhiser, R.E., Samuels, M.R., and Eliassen, J.D., *Chemical Engineering Thermodynamics, The Study of Energy, Entropy, and Equilibrium*, Prentice-Hall, Englewood Cliffs, NJ, 1972.

Dierdorf, D., personal communication, 1995.

Felder, P., "The influence of the molecular beam temperature on the photodissociation of CF_3I at 308 nm," *Chem. Phys. Lett.* **197**, 425 (1992).

Fourier Infrared Software Tools for Microsoft Windows, User's Reference Guide, Mattson Instruments, Inc., 1992.

McGee, P. R., Cleveland, F. F., Meister, A. G., Decker, C. E., and Miller, S. I., "Substituted Methanes.X. Infrared Spectral Data, Assignments, Potential Constants, and Calculated Thermodynamic Properties for CF_3Br and CF_3I," *J. of Chem. Phys.*, **21**, 2, 1948.

Mendenhall, W. and Sincich, T., *Statistics for Engineering and the Sciences*, Dellen Publishing Co., p. 213, 214, 1992.

Peacock, R.D., Cleary, T.G., and Harris, Jr., R.H., "Agent Stability under Storage and Discharge Residue," Section 6 in *Evaluation of Alternative In-Flight Fire Suppressants for Full-Scale Testing in Simulated Aircraft Engine Nacelles and Dry Bays*, NIST Special Publication 861, Grosshandler, W. L., Gann, R. G., and Pitts, W. M., eds, U.S. Department of Commerce, Washington, DC 1994. 643-668.

Ricker, R.E., Stoudt, M.R., Dante, J.F., Fink, J.L., Beauchamp, C.R., and Moffat, T.P., "Corrosion of Metals," Section 7 in *Evaluation of Alternative In-Flight Fire Suppressants for Full-Scale Testing in Simulated Aircraft Engine Nacelles and Dry Bays*, NIST Special Publication 861, Grosshandler, W. L., Gann, R. G., and Pitts, W. M., eds, U.S. Department of Commerce, Washington, DC 1994. 643-668.

Taylor, B.N., Kuyatt, C.E., *Guidelines for Evaluating and Expressing the Uncertainty of NIST Measurement Results*, NIST Technical Note 1297, U.S. Department of Commerce, Washington DC. 1994.

Appendix A. Initial and Final FTIR Spectra for HFC-125

The spectra in Appendix A are those of the HFC-125. The gas cell pressure for all was 5330 Pa. The lower spectrum is for the initial analysis; the upper spectrum is for the final aged analysis. The blanks for this agent appear in Figures 13 and 14.

7. AGENT STABILITY UNDER STORAGE

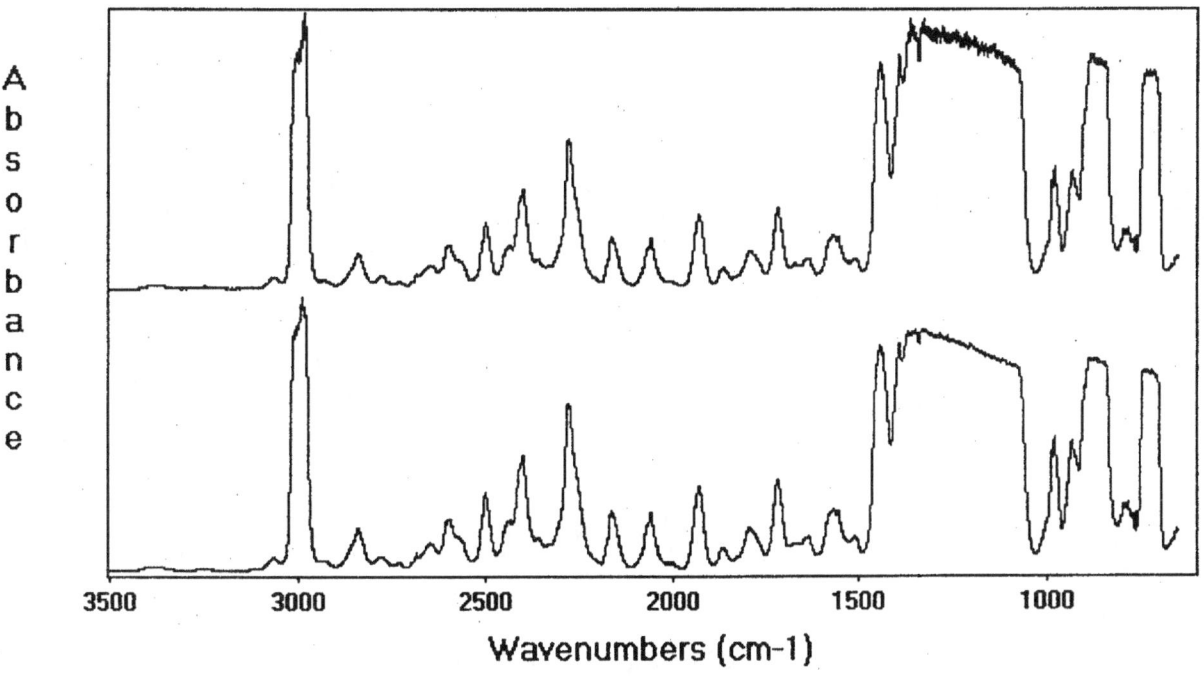

Figure A-1. Initial (lower) and 48 week (upper) spectra for Nitronic 40 in HFC-125 at 23 °C.

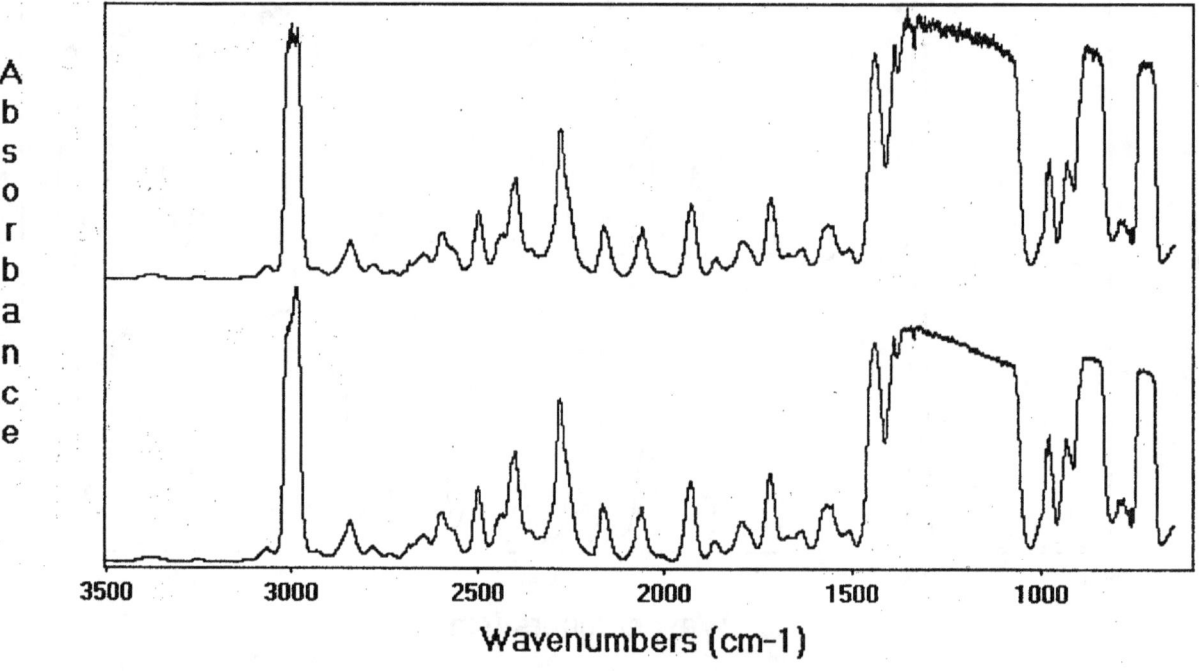

Figure A-2.   Initial (lower) and 48 week (upper) spectra for Ti-15-3-3-3 in HFC-125 at 23 °C.

7. AGENT STABILITY UNDER STORAGE

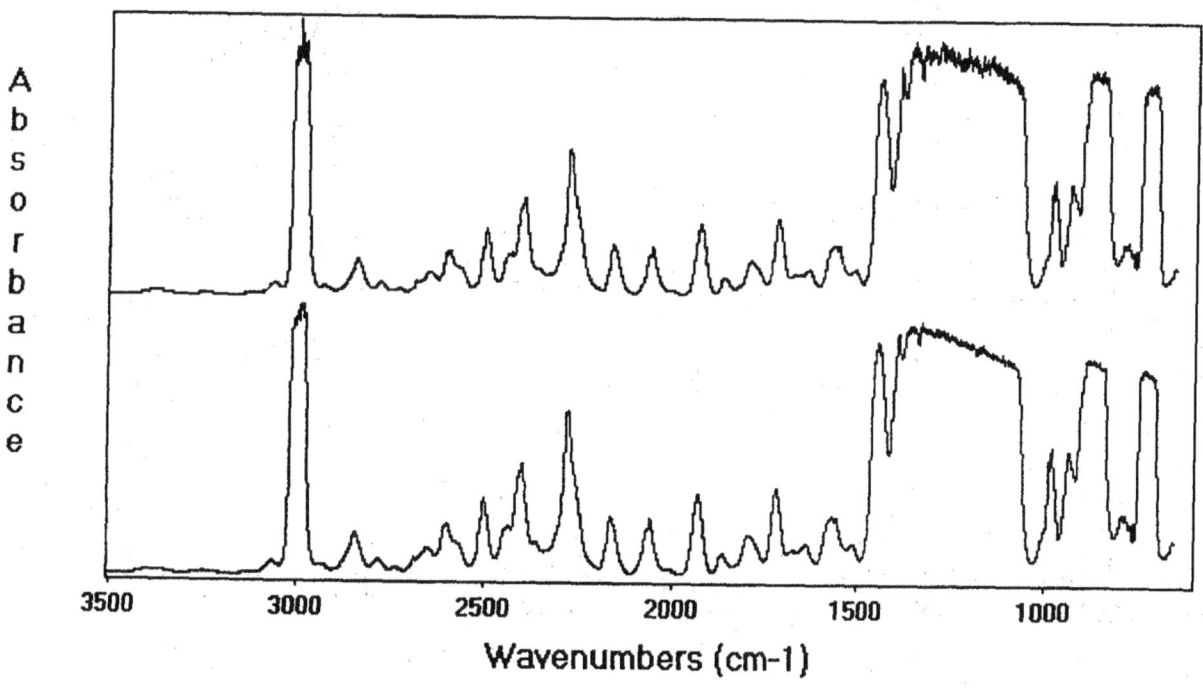

Figure A-3.   Initial (lower) and 32 week (upper) spectra for C4130 in HFC-125 at 23 °C.

344 7. AGENT STABILITY UNDER STORAGE

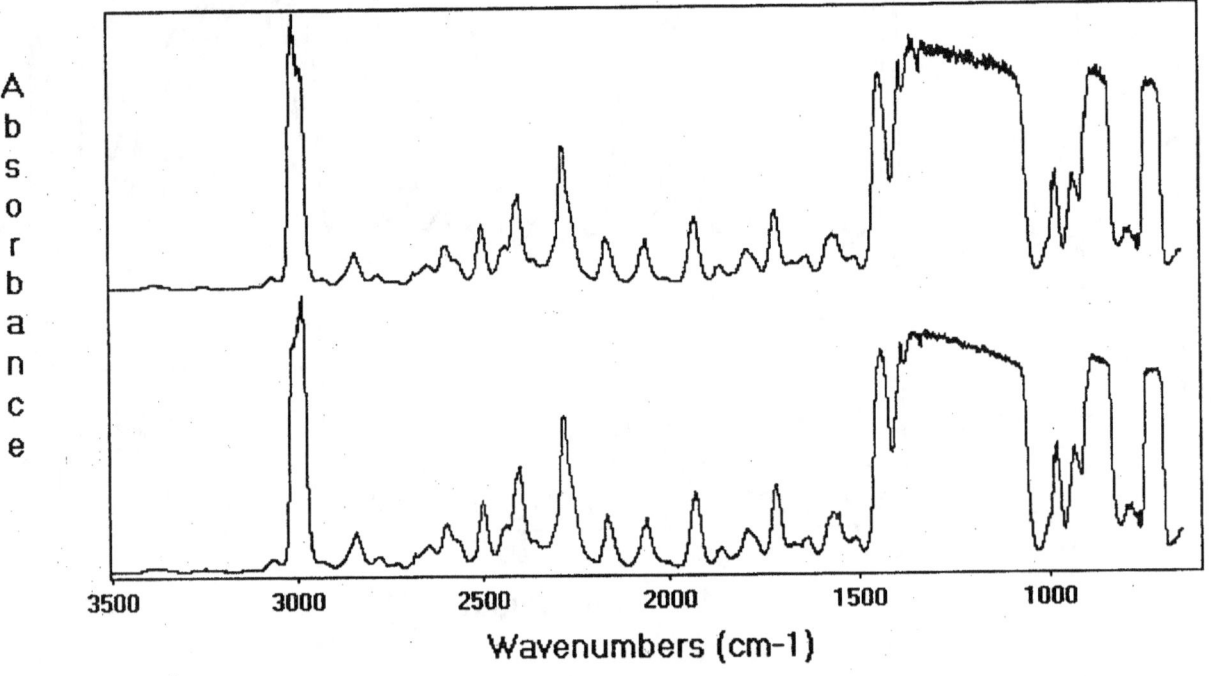

Figure A-4. Initial (lower) and 48 week (upper) spectra for I625 in HFC-125 at 23 °C.

7. AGENT STABILITY UNDER STORAGE

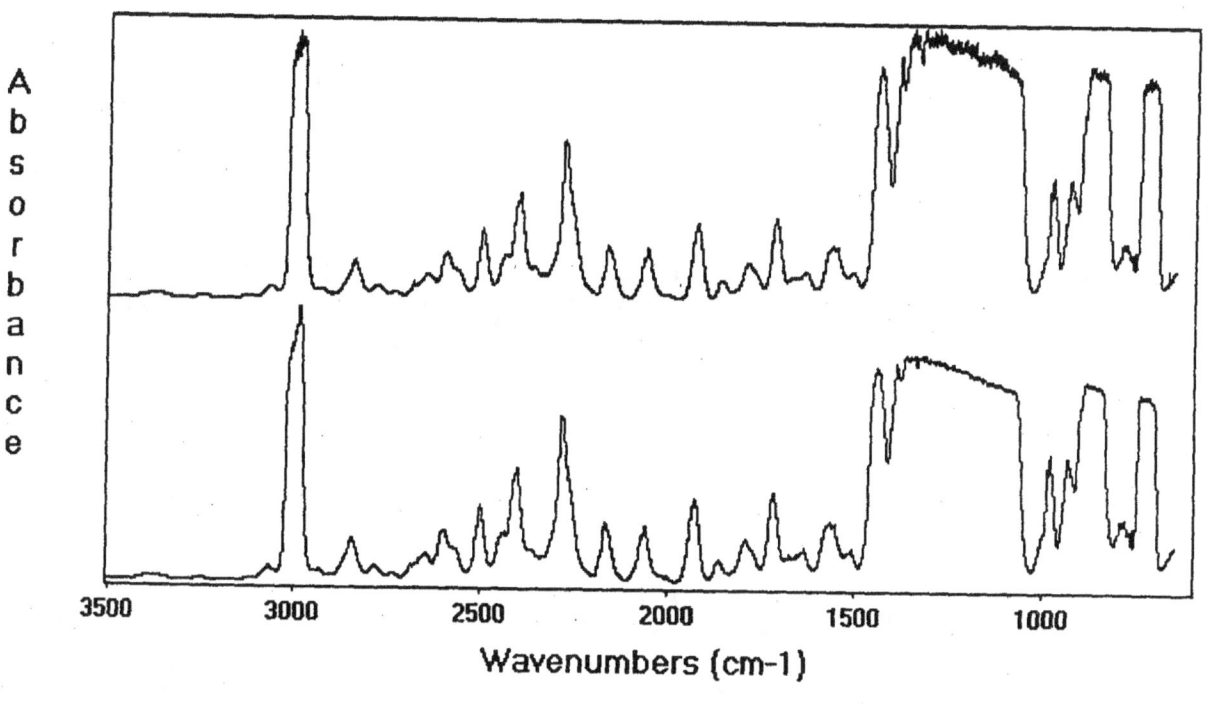

Figure A-5.   Initial (lower) and 48 week (upper) spectra for nitronic 40 in HFC-125 at 150 °C.

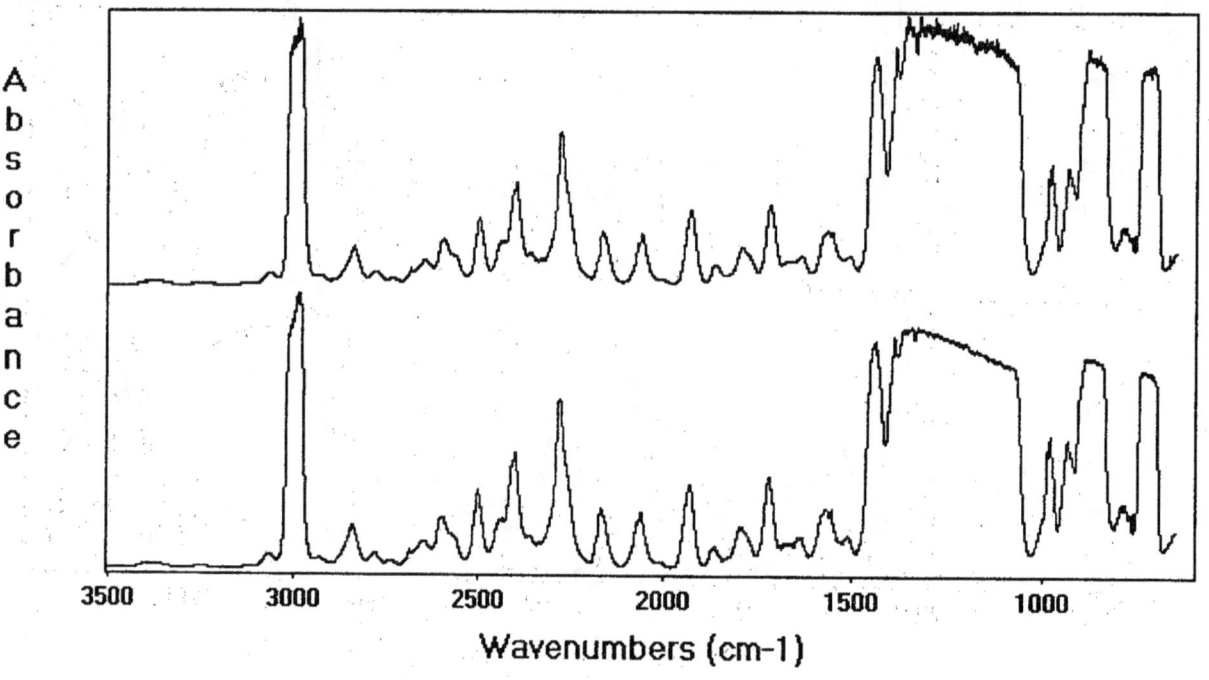

Figure A-6.   Initial (lower) and 48 week (upper) spectra for Ti-15-3-3-3 in HFC-125 at 150 °C.

7. AGENT STABILITY UNDER STORAGE

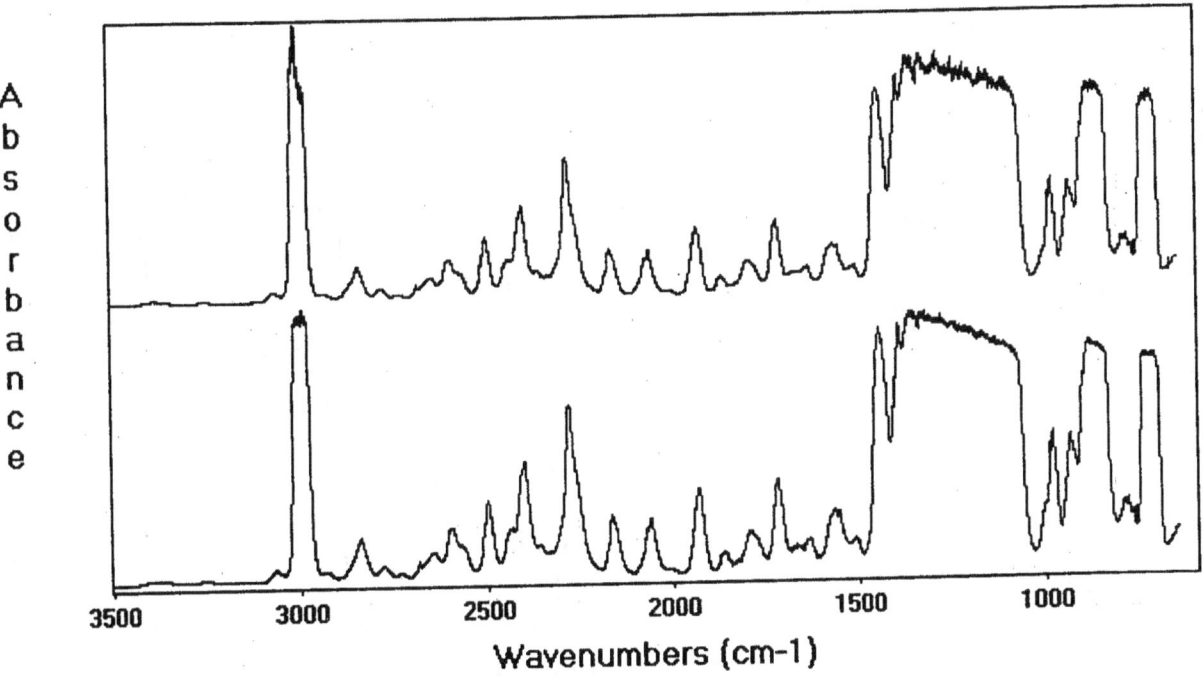

Figure A-7. Initial (lower) and 40 week (upper) spectra for C4130 in HFC-125 at 150 °C.

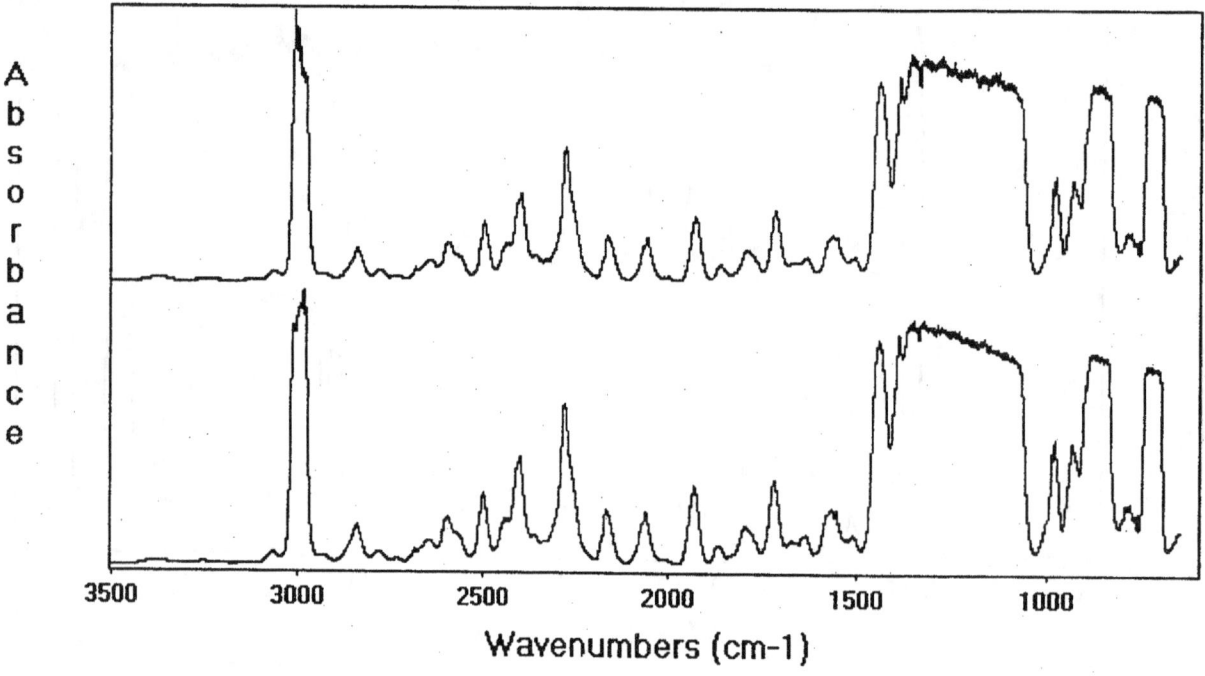

Figure A-8. Initial (lower) and 48 week (upper) spectra for I625 in HFC-125 at 150 °C.

7. AGENT STABILITY UNDER STORAGE 349

Appendix B. Initial and Final FTIR Spectra of HFC-227ea

The spectra in Appendix B are those of the HFC-227ea. The gas cell pressure for all was 5330 Pa. The lower spectrum is for the initial analysis; the upper spectrum is for the final aged analysis. The blanks for this agent appear in Figures 17-19.

350 7. AGENT STABILITY UNDER STORAGE

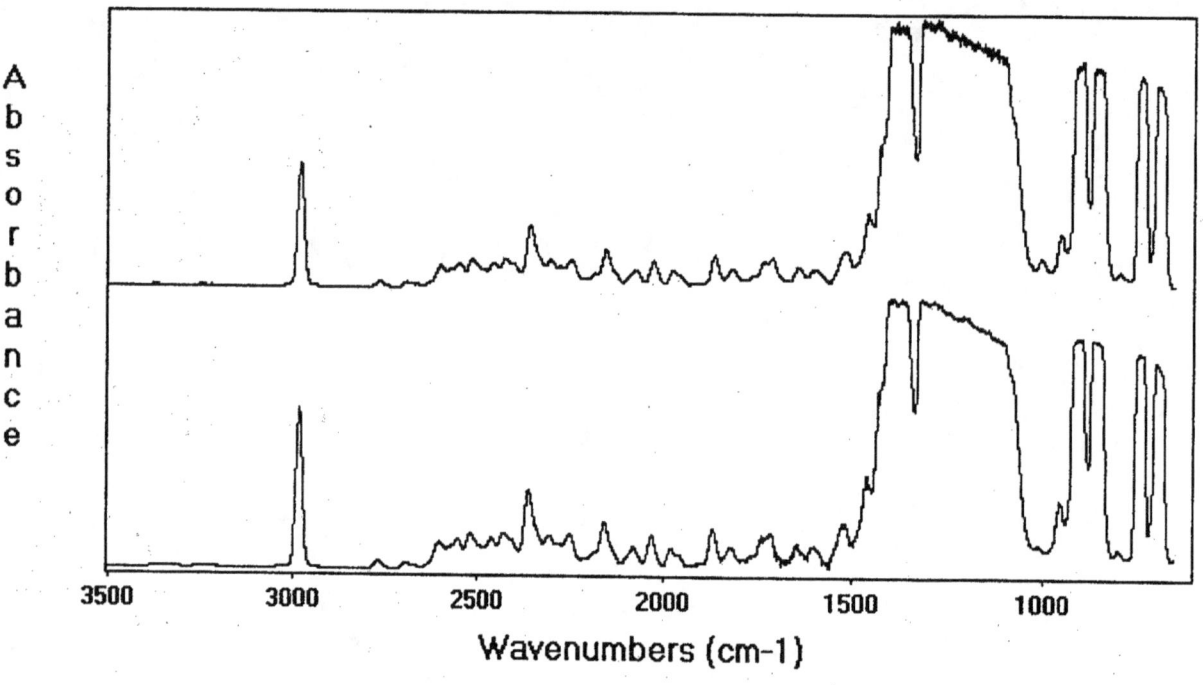

Figure B-1. Initial (lower) and 48 week (upper) spectra for nitronic 40 in HFC-227ea at 23 °C.

7. AGENT STABILITY UNDER STORAGE

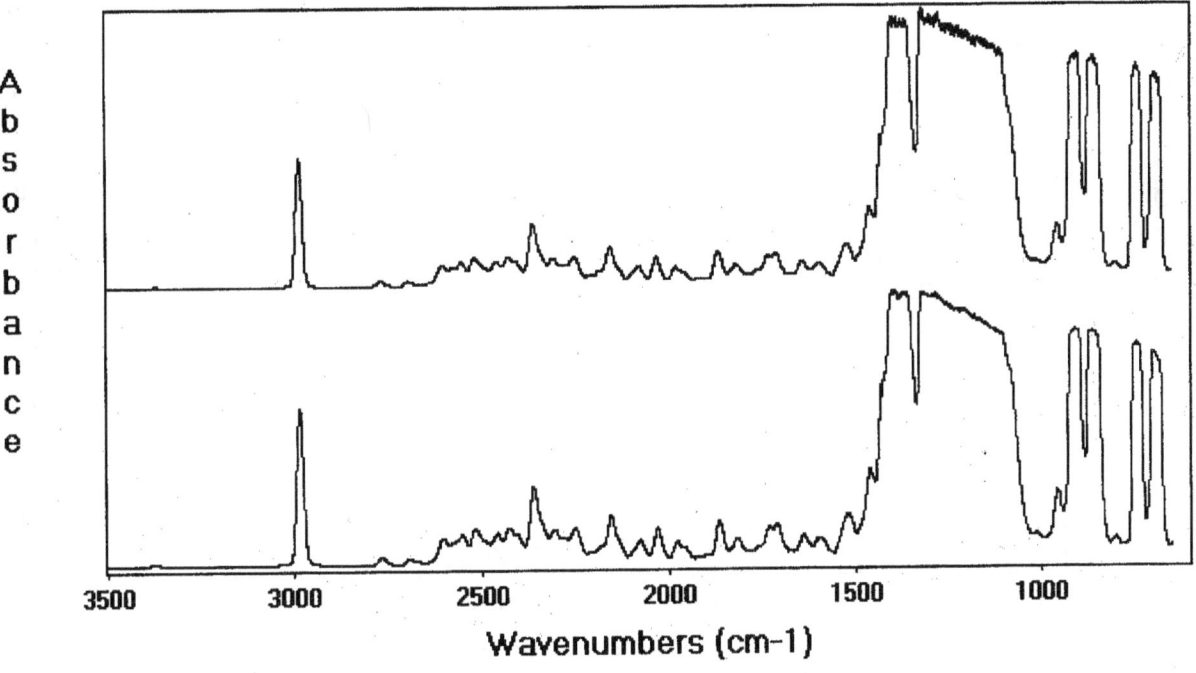

Figure B-2. Initial (lower) and 40 week (upper) spectra for Ti-15-3-3-3 in HFC-227ea at 23 °C.

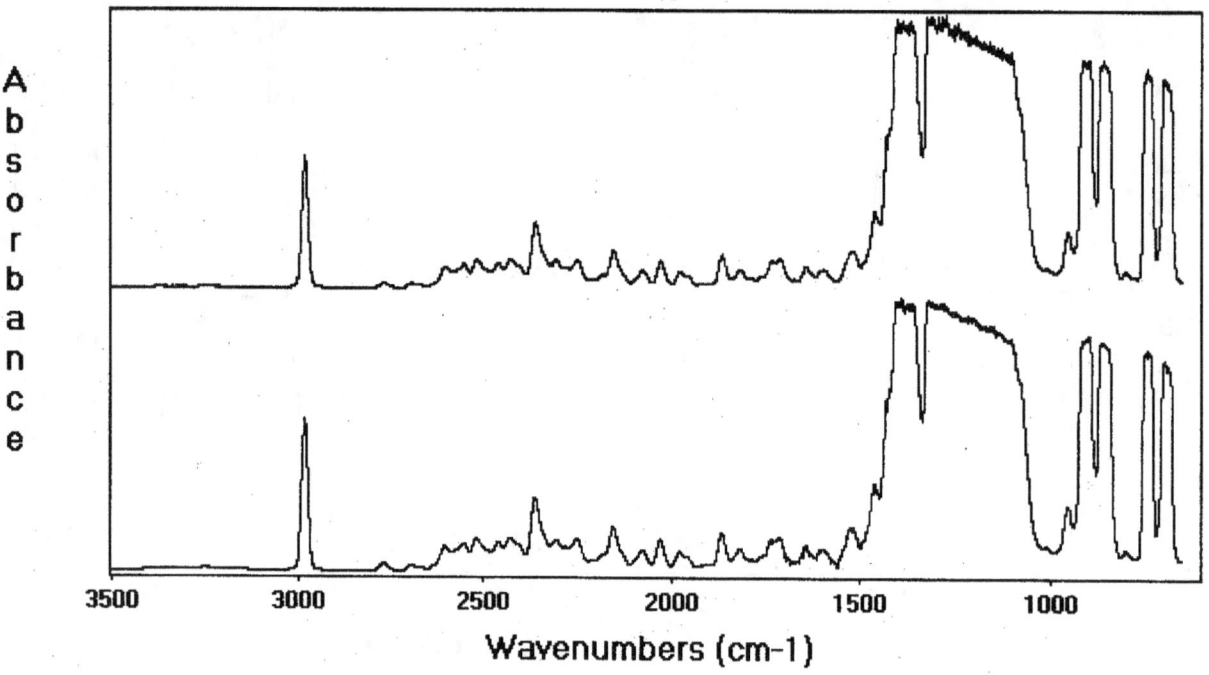

Figure B-3. Initial (lower) and 32 week (upper) spectra for C4130 in HFC-227ea at 23 °C.

7. AGENT STABILITY UNDER STORAGE

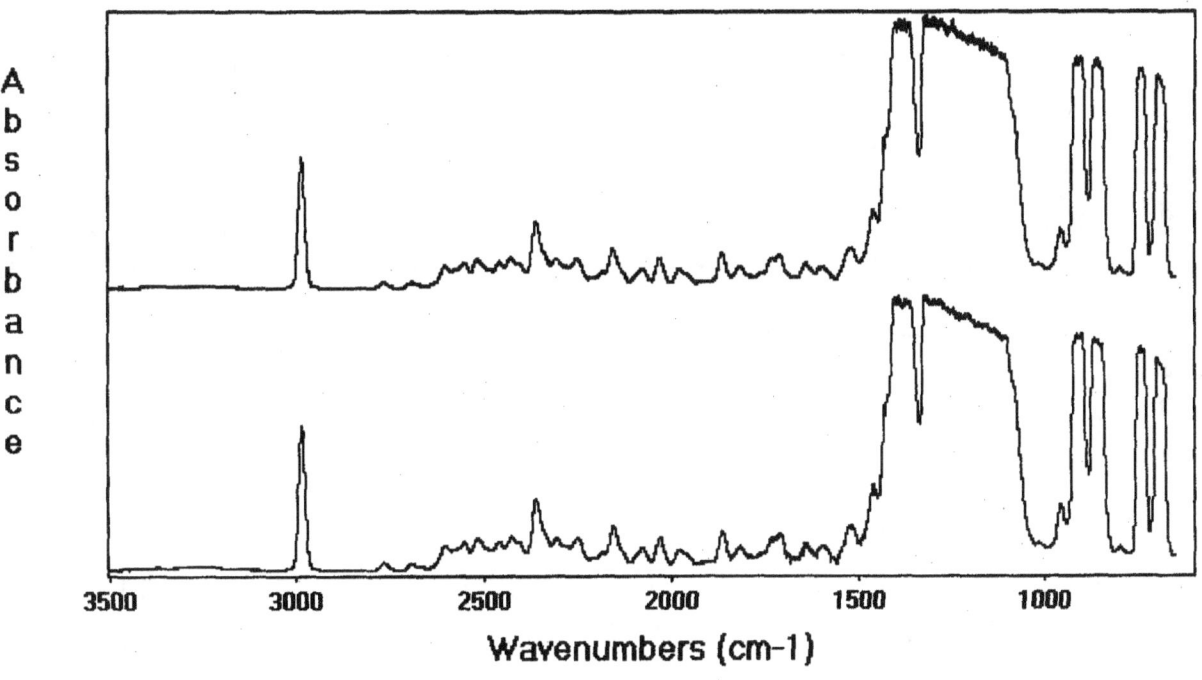

Figure B-4. Initial (lower) and 48 week (upper) spectra for I625 in HFC-227ea at 23 °C.

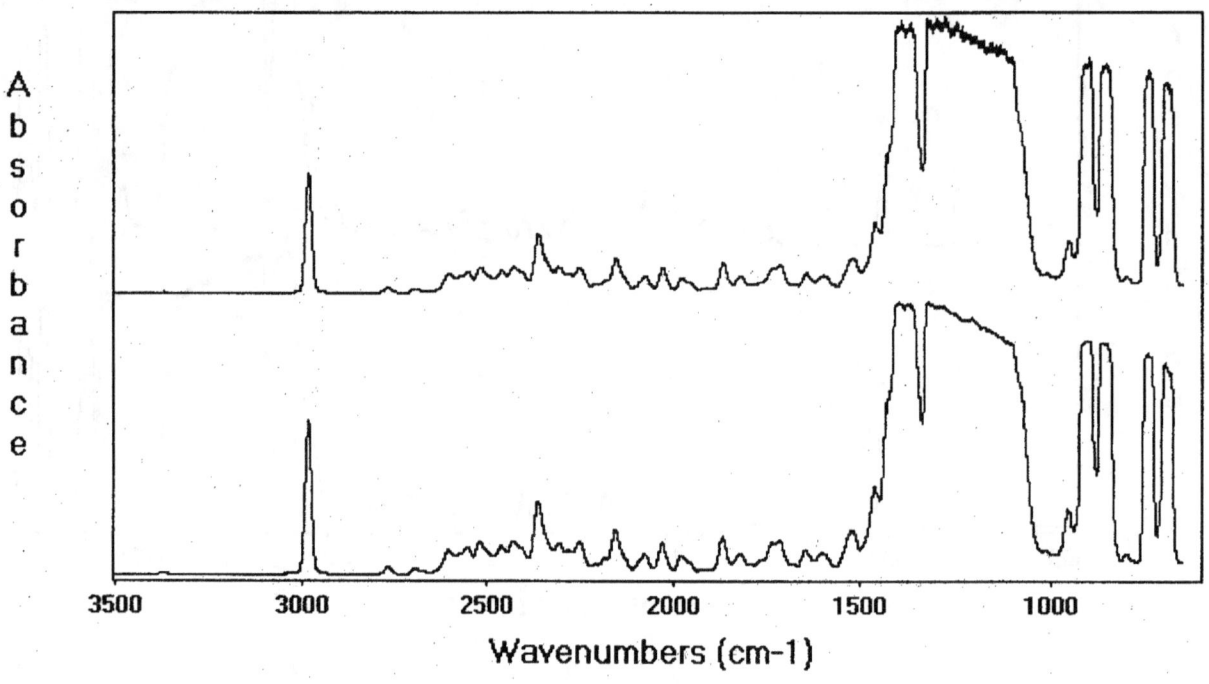

Figure B-5. Initial (lower) and 40 week (upper) spectra for nitronic 40 in HFC-227ea at 125 °C.

7. AGENT STABILITY UNDER STORAGE

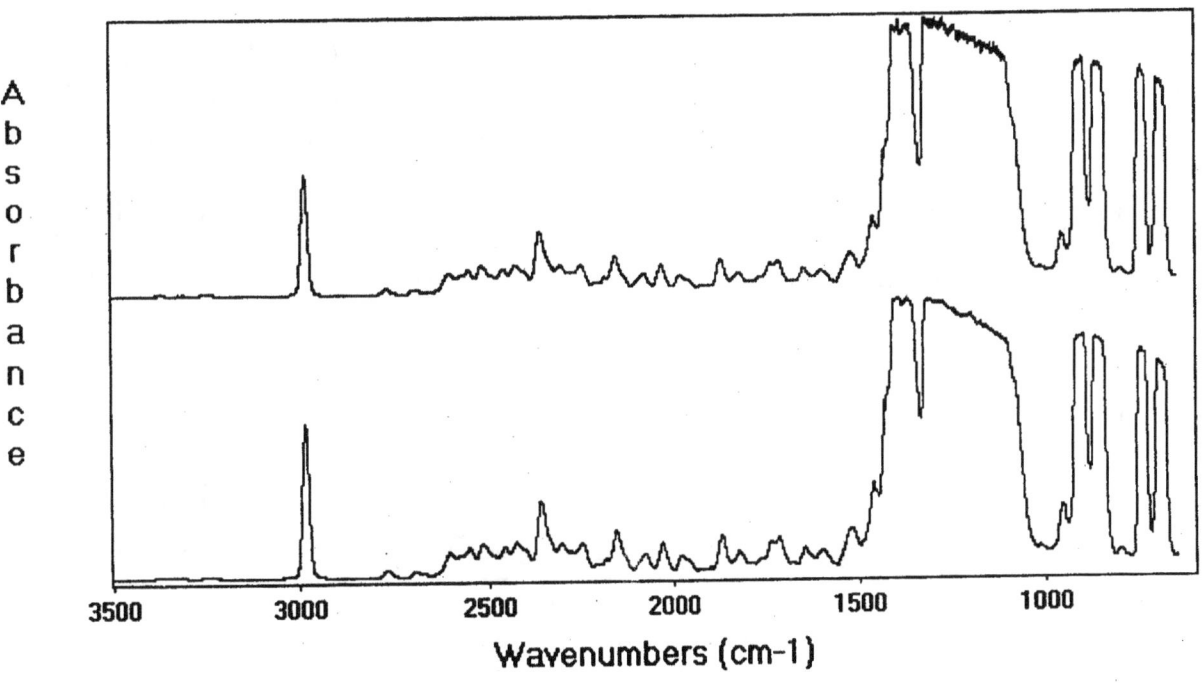

Figure B-6. Initial (lower) and 40 week (upper) spectra for Ti-15-3-3-3 in HFC-227ea at 125 °C.

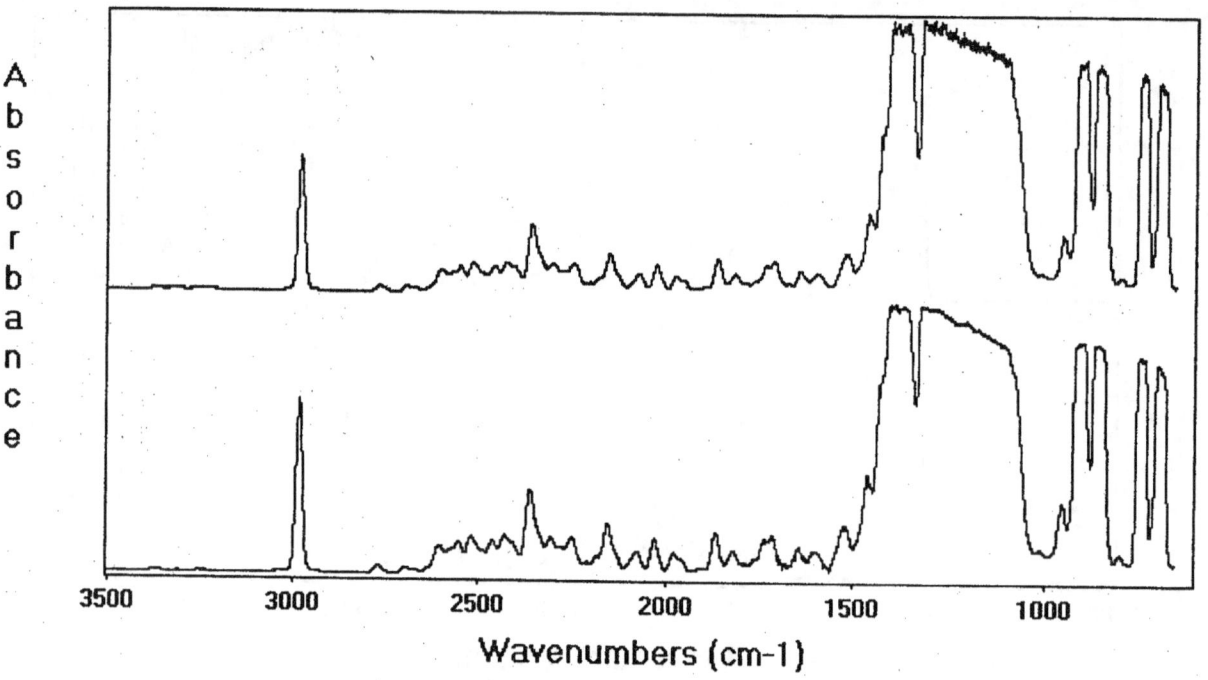

Figure B-7. Initial (lower) and 40 week (upper) spectra for C4130 in HFC-227ea at 125 °C.

7. AGENT STABILITY UNDER STORAGE

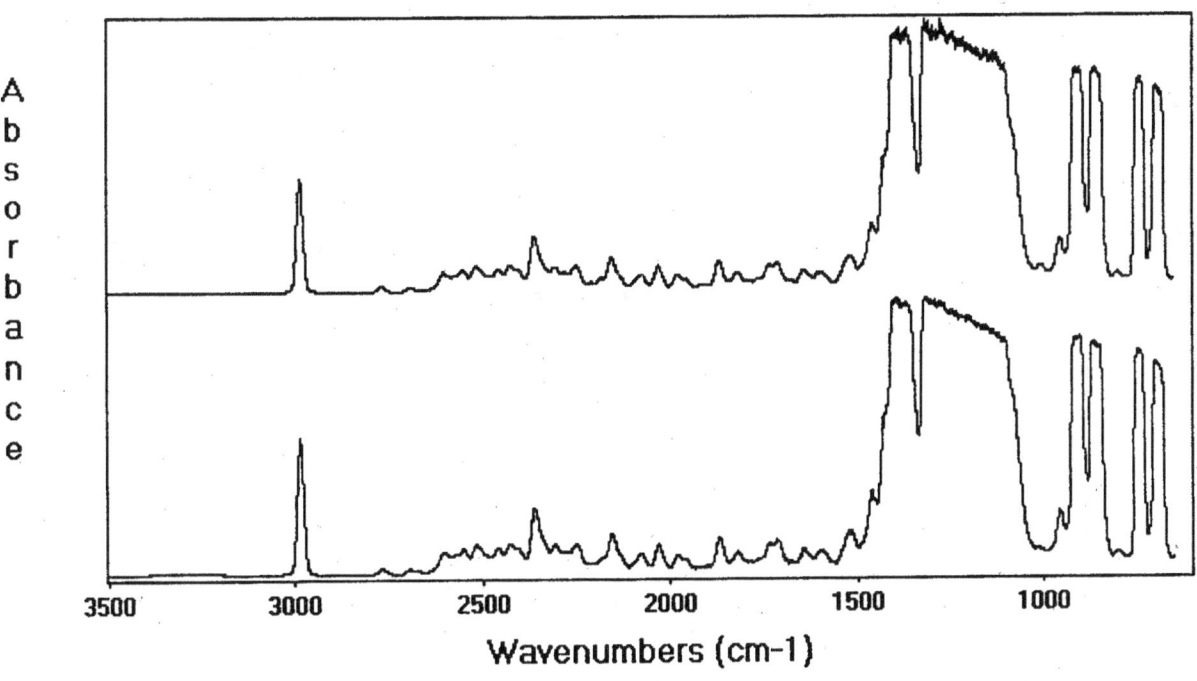

Figure B-8. Initial (lower) and 40 week (upper) spectra for I625 in HFC-227ea at 125 °C.

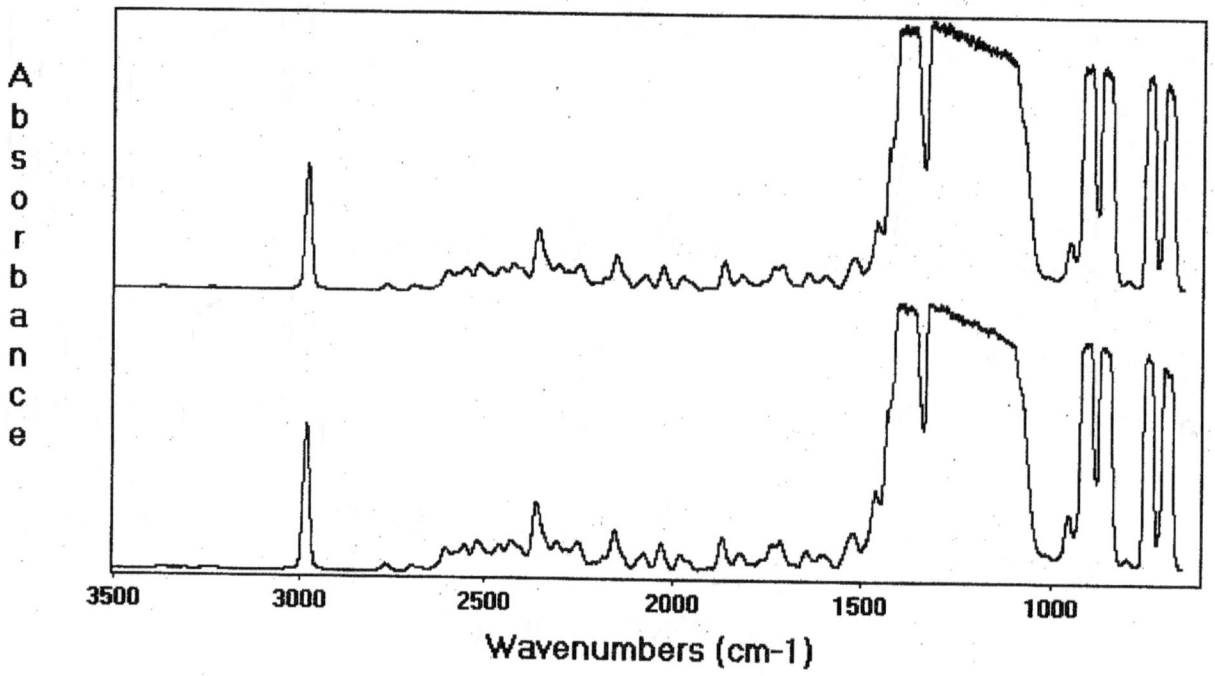

Figure B-9. Initial (lower) and 48 week (upper) spectra for nitronic 40 in HFC-227ea at 150 °C.

7. AGENT STABILITY UNDER STORAGE

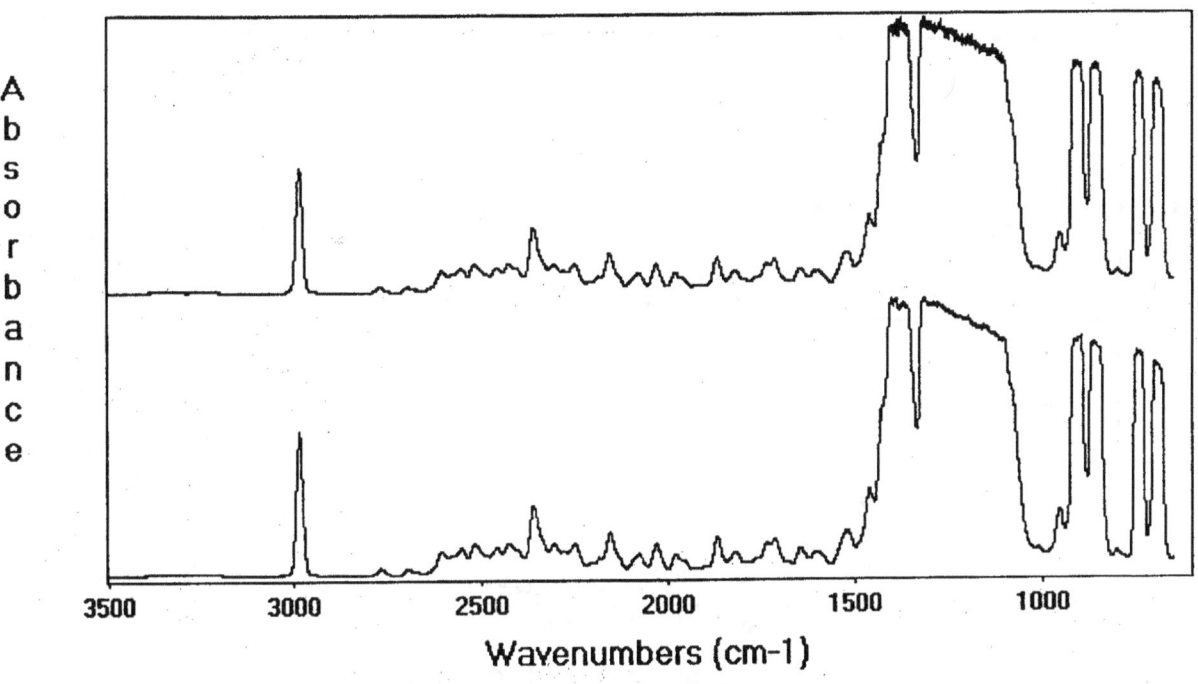

Figure B-10. Initial (lower) and 48 week (upper) spectra for Ti-15-3-3-3 in HFC-227ea at 150 °C.

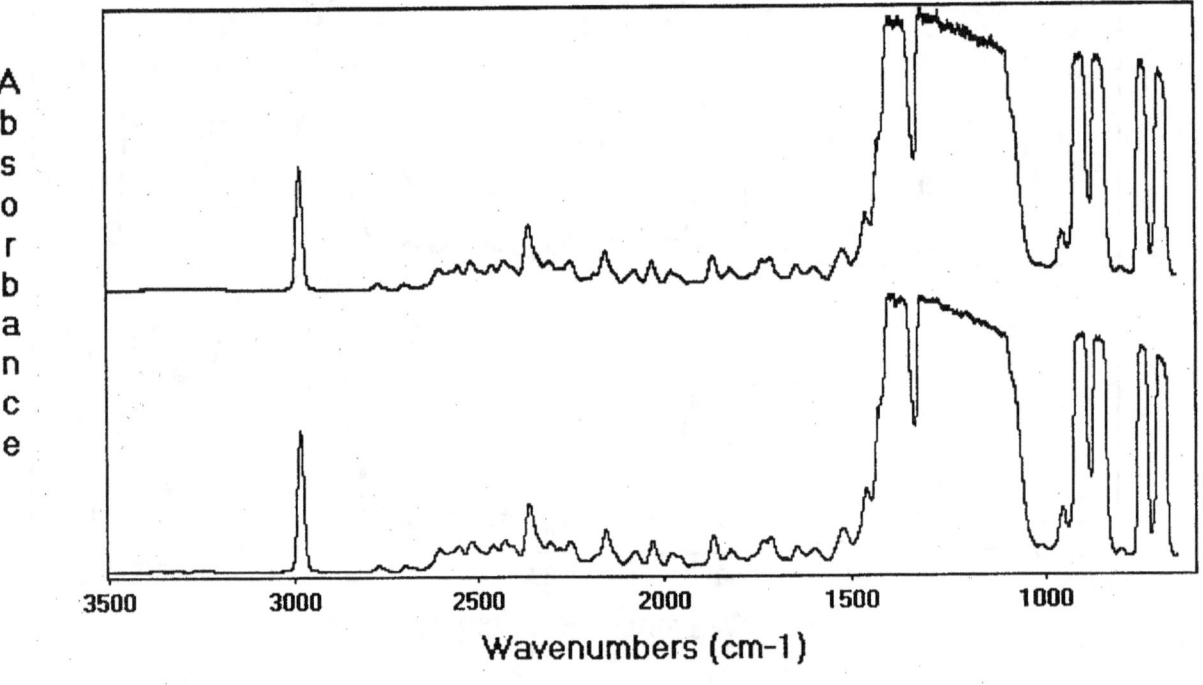

Figure B-11. Initial (lower) and 40 week (upper) spectra for C4130 in HFC-227ea at 150 °C.

7. AGENT STABILITY UNDER STORAGE

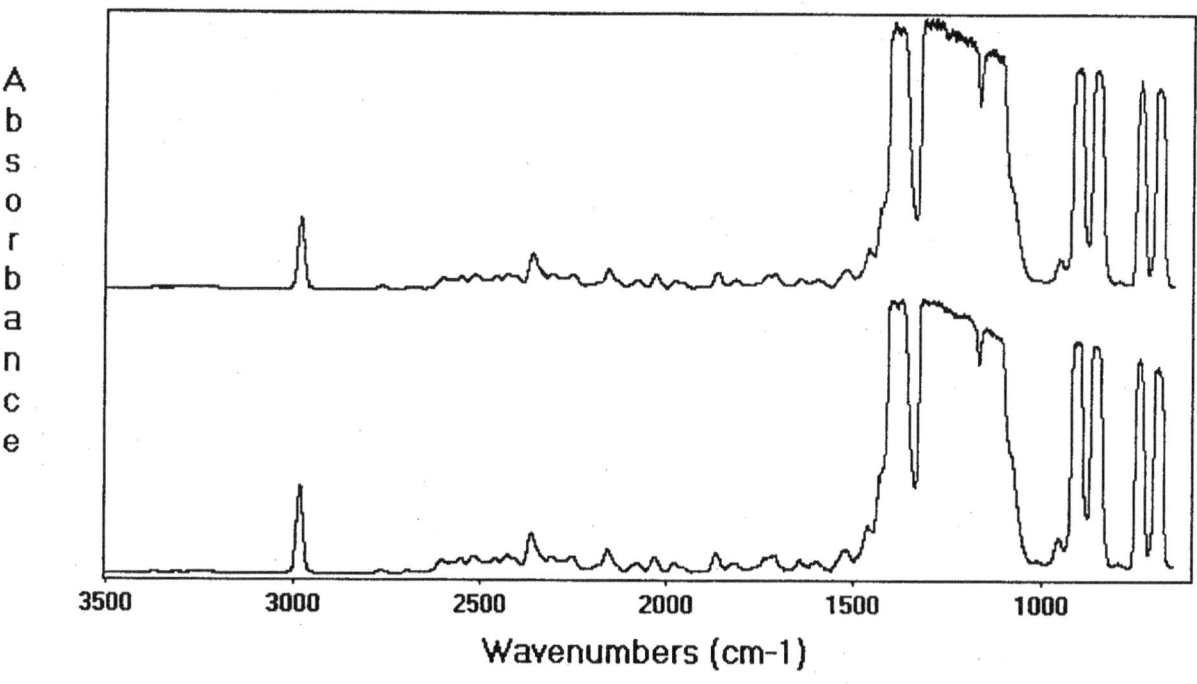

Figure B-12. Initial (lower) and 48 week (upper) spectra for I625 in HFC-227ea at 150 °C.

Appendix C. Initial and Final FTIR Spectra of CF_3I Tested in the Dry Condition without Copper

The spectra in Appendix C are those of the CF_3I. The gas cell pressure for all was 5330 Pa. The lower spectrum is for the initial analysis; the upper spectrum is for the final aged analysis.

7. AGENT STABILITY UNDER STORAGE

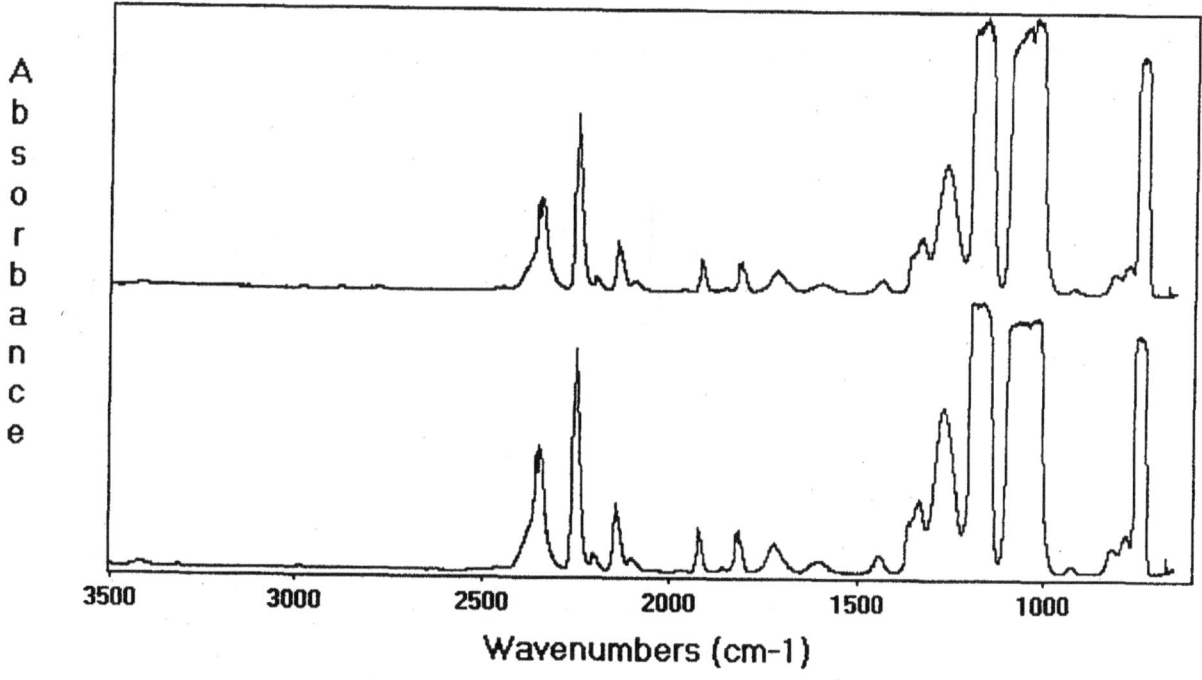

Figure C-1. Initial (lower) and 48 week (upper) spectra for the blank in CF_3I tested in the dry condition without copper at 23 °C.

364   7. AGENT STABILITY UNDER STORAGE

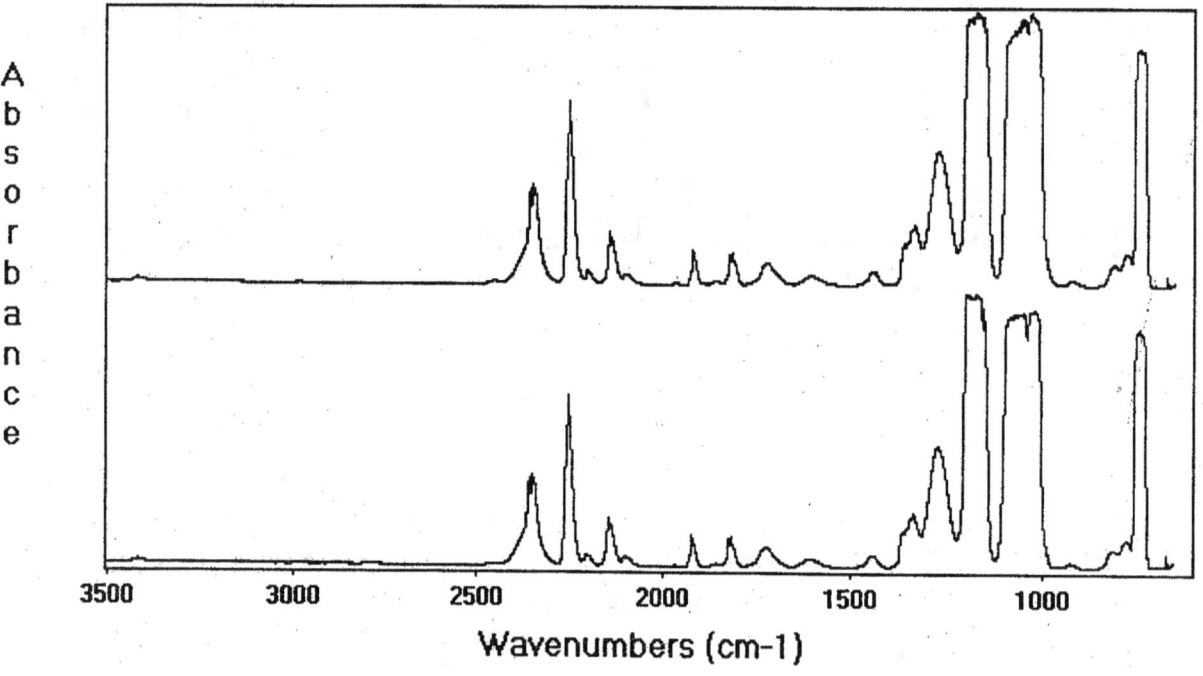

Figure C-2. Initial (lower) and 48 week (upper) spectra for nitronic 40 in CF_3I tested in the dry condition without copper at 23 °C.

7. AGENT STABILITY UNDER STORAGE

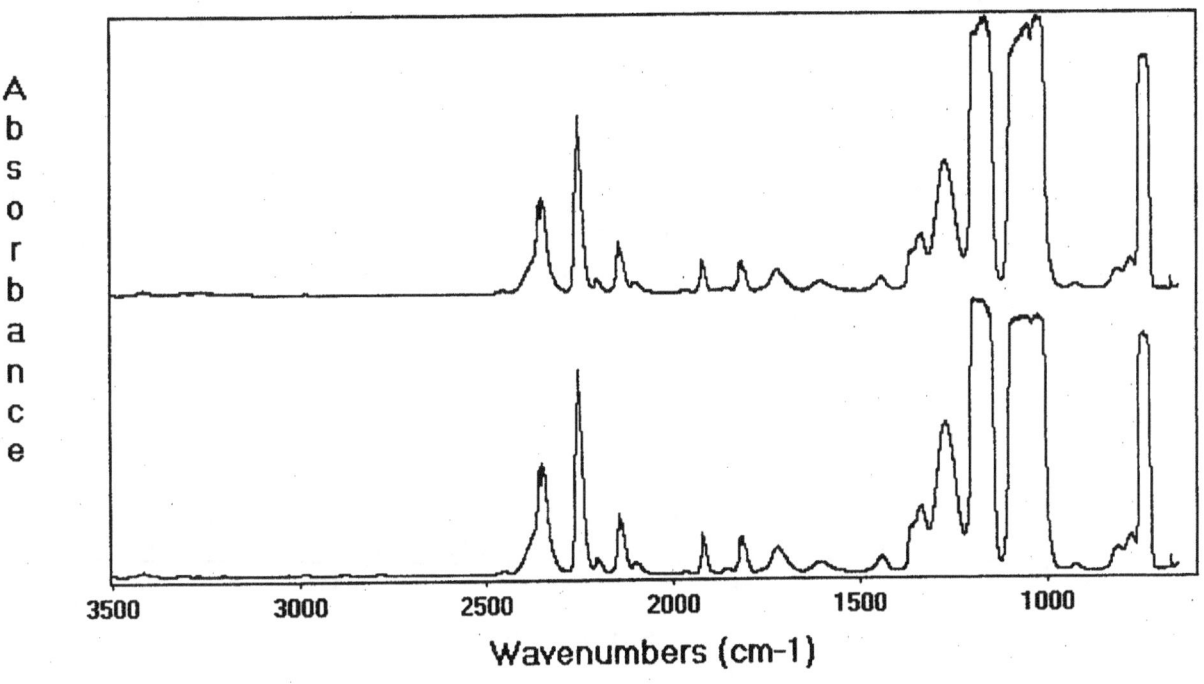

Figure C-3. Initial (lower) and 48 week (upper) spectra for Ti-15-3-3-3 in CF_3I tested in the dry condition without copper at 23 °C.

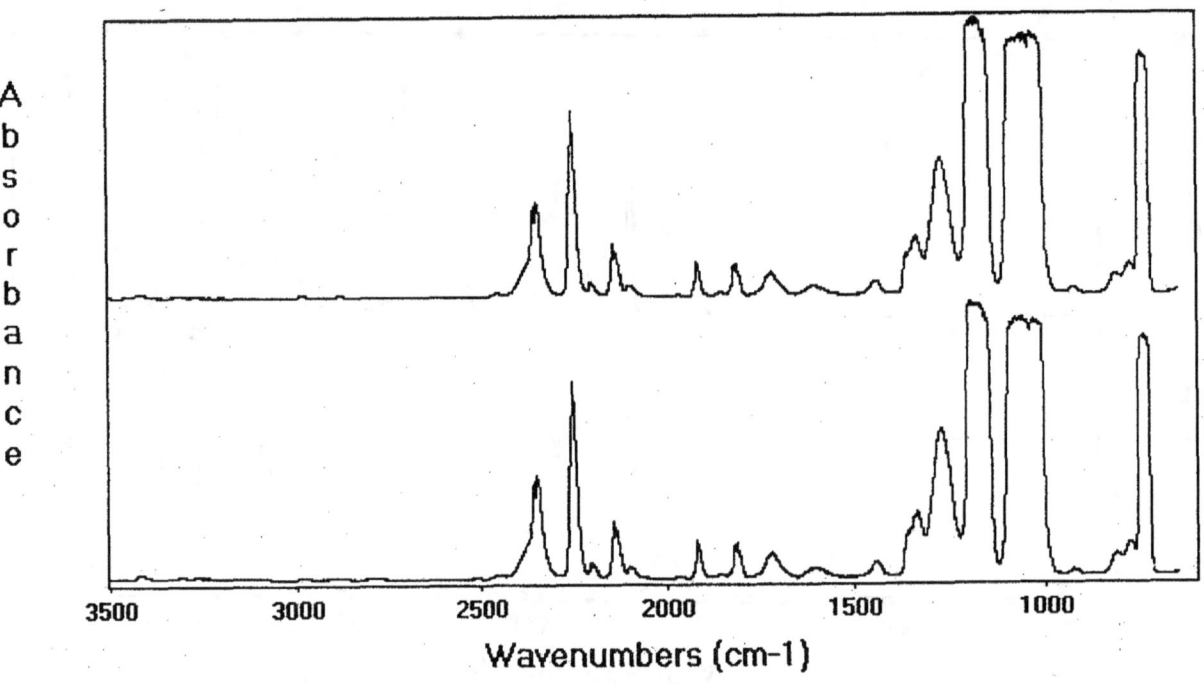

Figure C-4. Initial (lower) and 40 week (upper) spectra for C4130 in CF_3I tested in the dry condition without copper at 23 °C.

7. AGENT STABILITY UNDER STORAGE

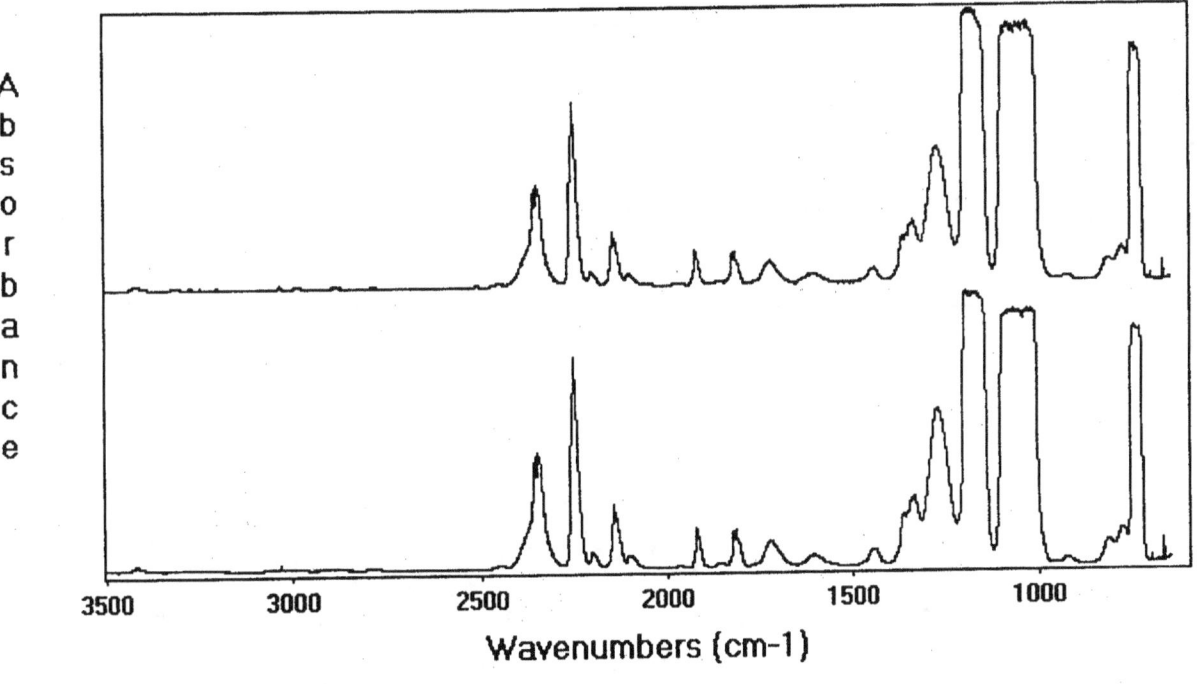

Figure C-5. Initial (lower) and 44 week (upper) spectra for I625 in CF_3I tested in the dry condition without copper at 23 °C.

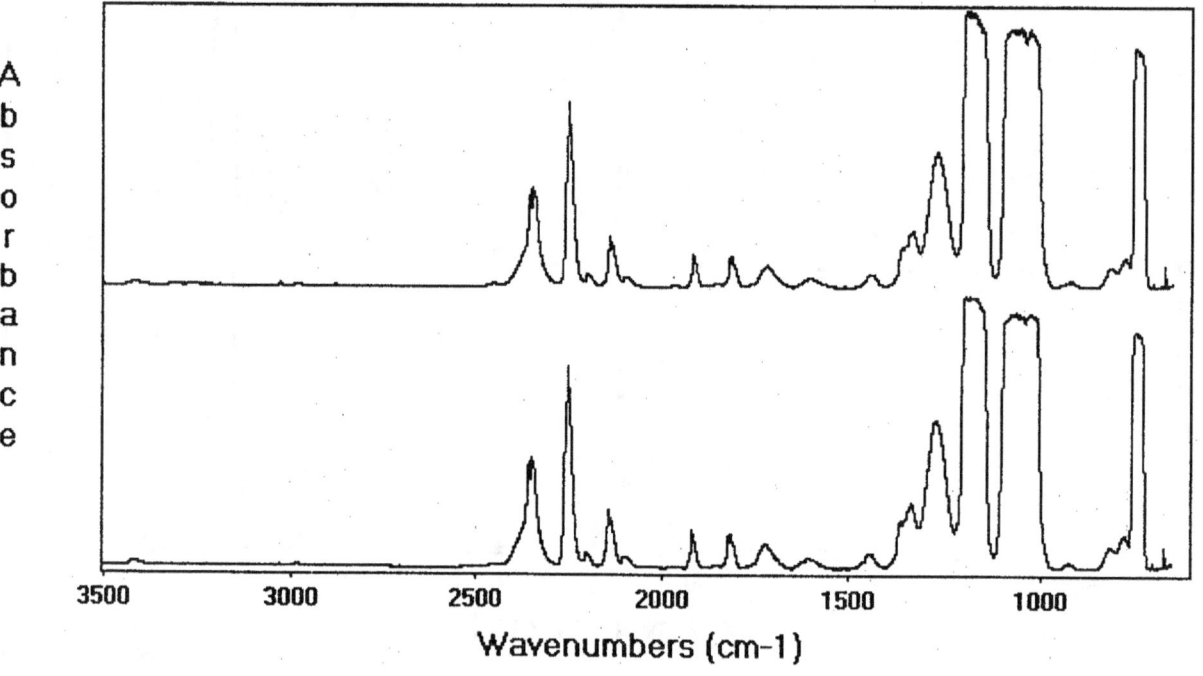

Figure C-6. Initial (lower) and 44 week (upper) spectra for the blank in CF_3I tested in the dry condition without copper at 100 °C.

7. AGENT STABILITY UNDER STORAGE

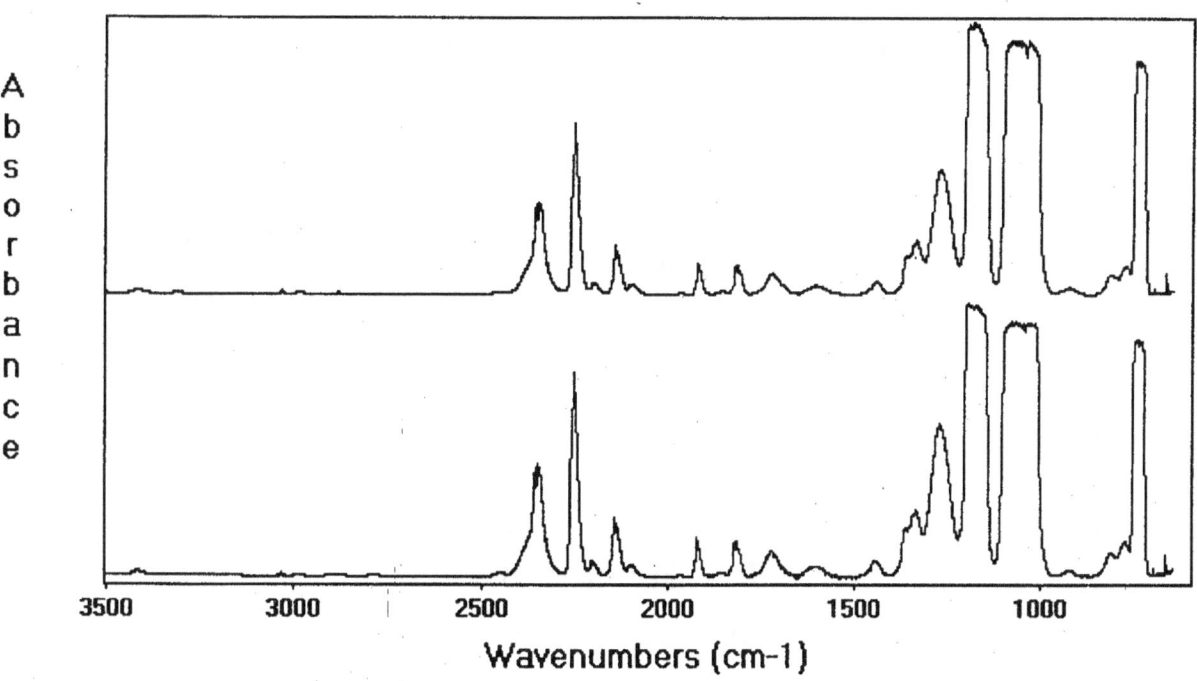

Figure C-7. Initial (lower) and 32 week (upper) spectra for nitronic 40 in CF_3I tested in the dry condition without copper at 100 °C.

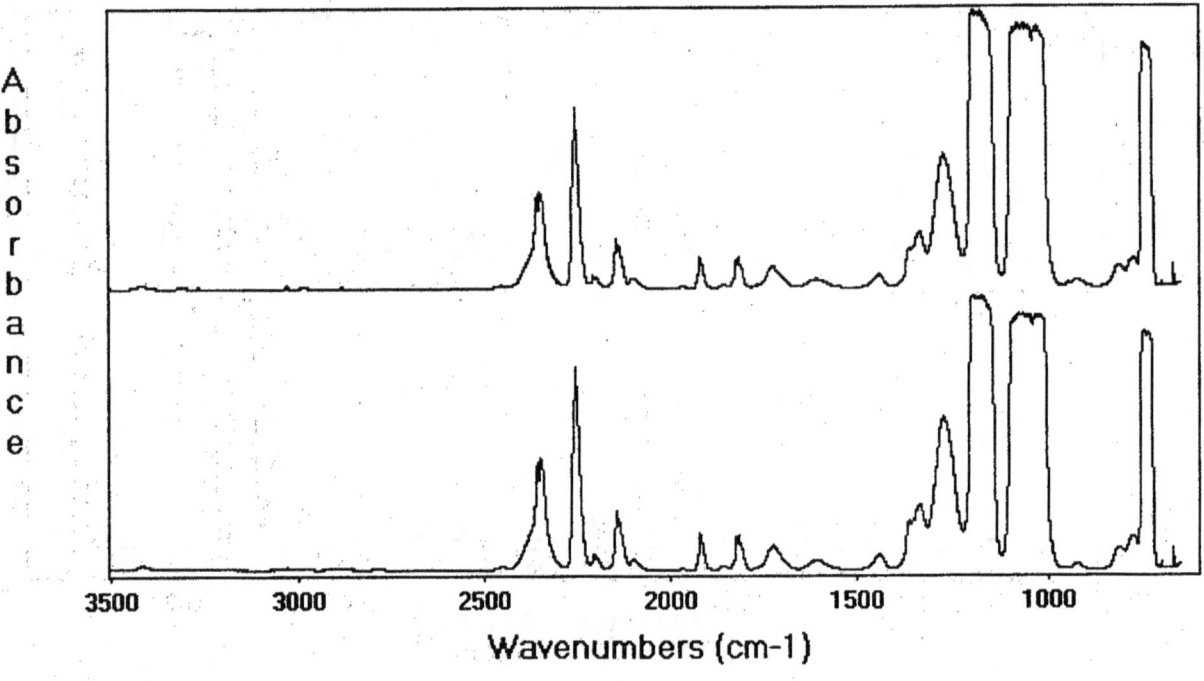

Figure C-8. Initial (lower) and 44 week (upper) spectra for Ti-15-3-3-3 in CF_3I tested in the dry condition without copper at 100 °C.

7. AGENT STABILITY UNDER STORAGE

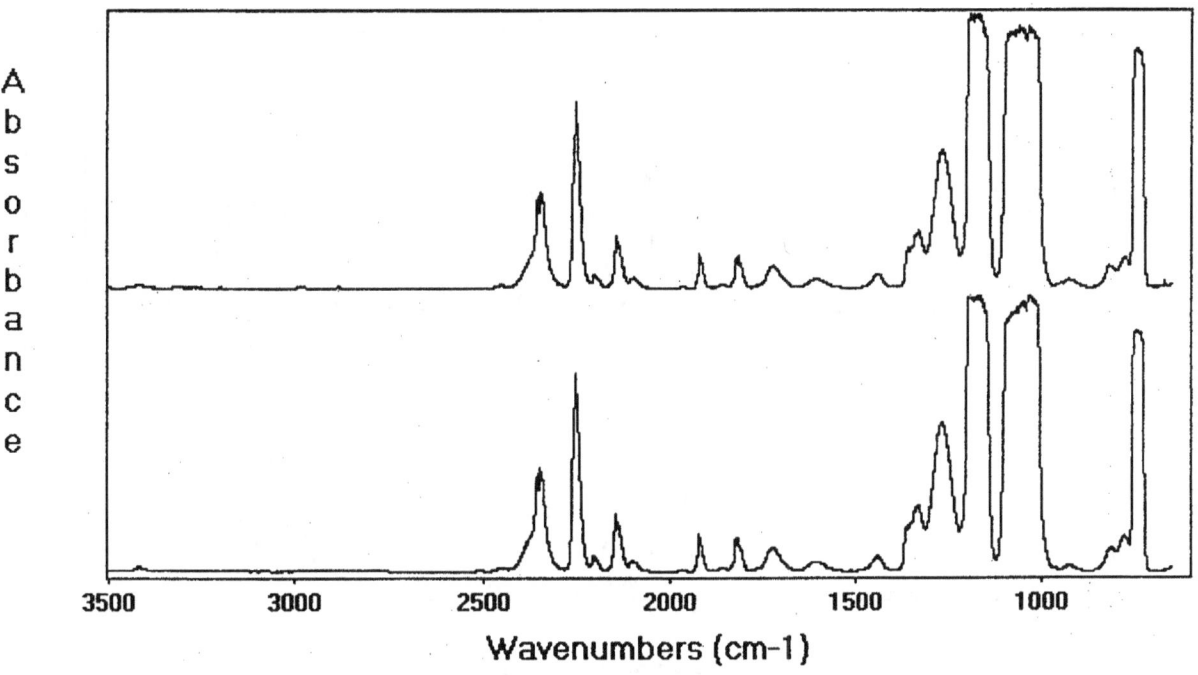

Figure C-9. Initial (lower) and 40 week (upper) spectra for C4130 in CF_3I tested in the dry condition without copper at 100 °C.

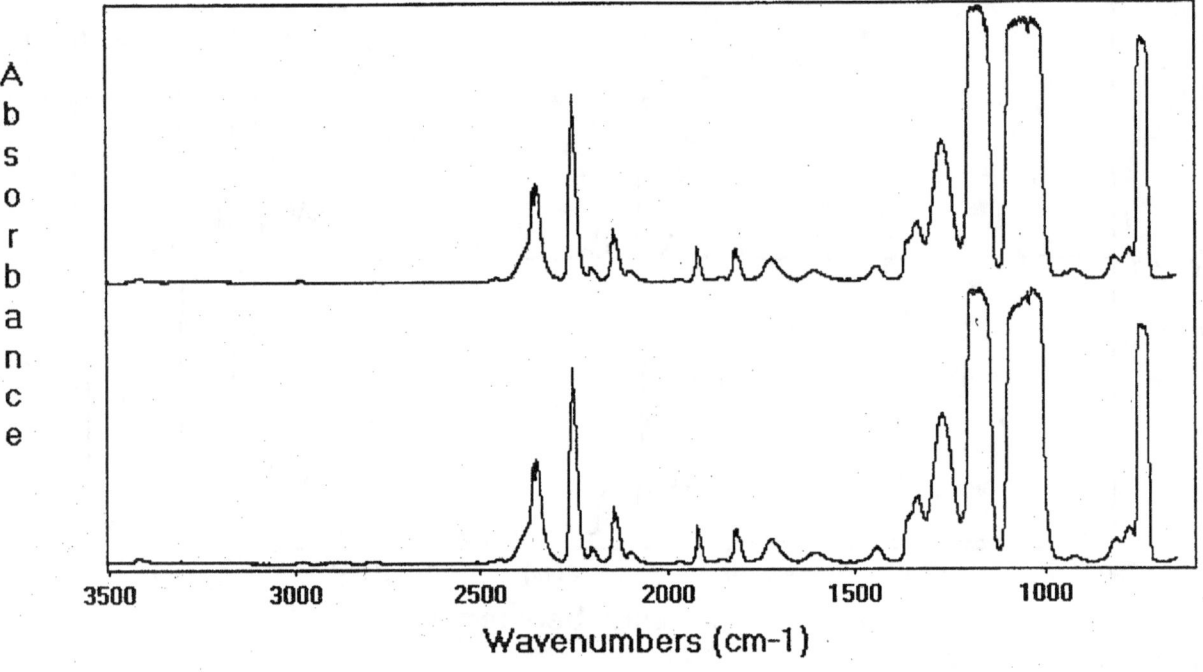

Figure C-10. Initial (lower) and 40 week (upper) spectra for I625 in CF_3I tested in the dry condition without copper at 100 °C.

7. AGENT STABILITY UNDER STORAGE

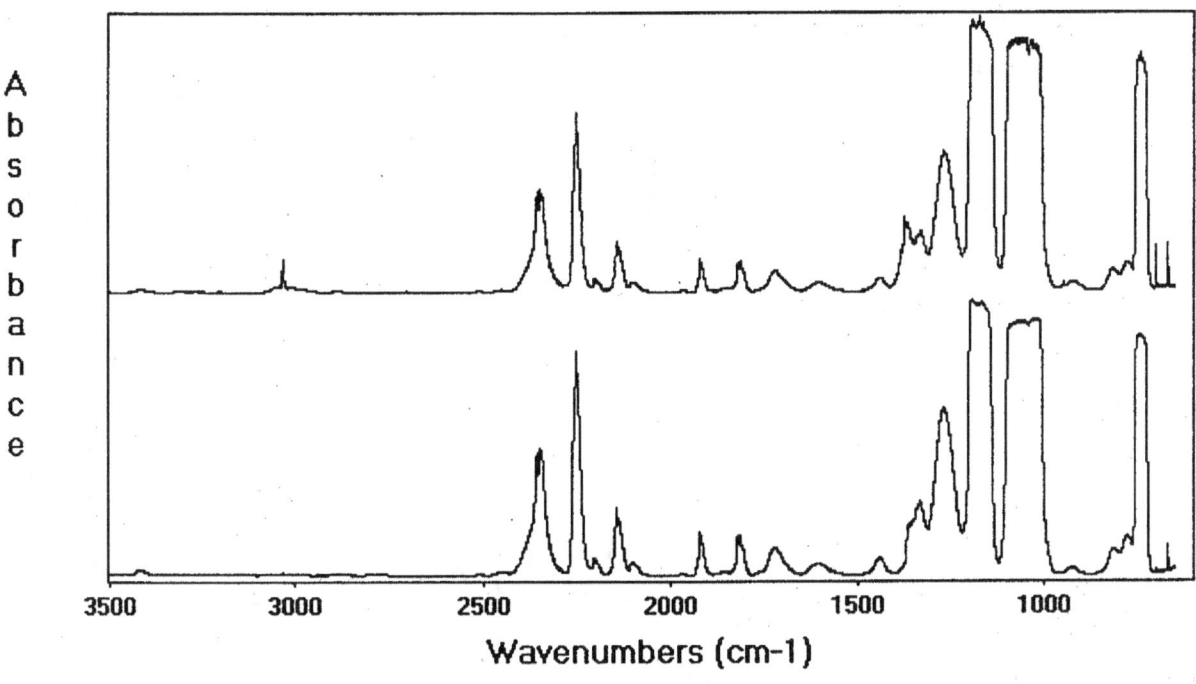

Figure C-11. Initial (lower) and 52 week (upper) spectra for the blank in CF_3I tested in the dry condition without copper at 150 °C.

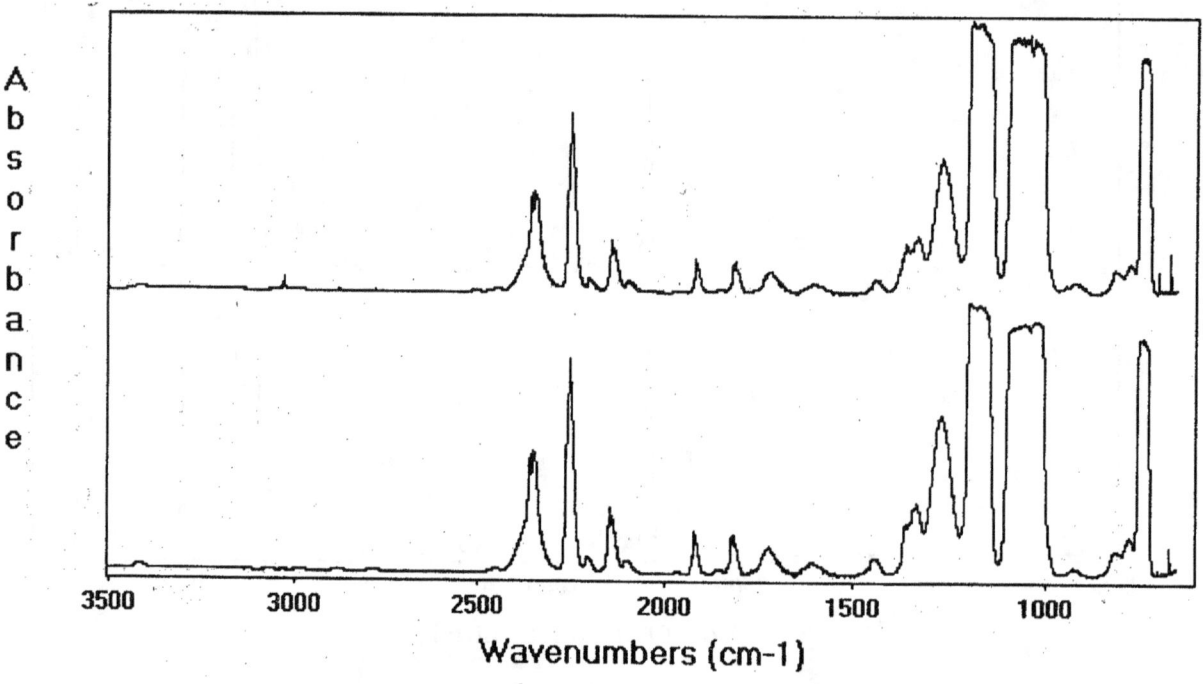

Figure C-12. Initial (lower) and 52 week (upper) spectra for nitronic 40 in CF$_3$I tested in the dry condition without copper at 150 °C.

7. AGENT STABILITY UNDER STORAGE

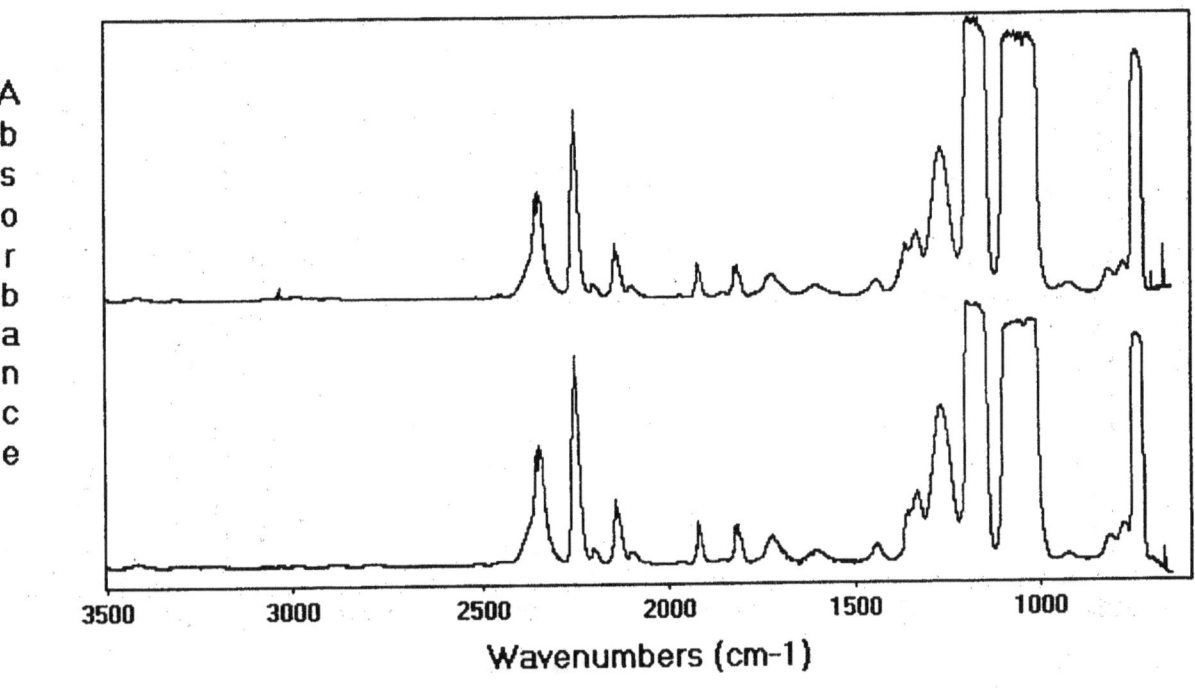

Figure C-13. Initial (lower) and 52 week (upper) spectra for Ti-15-3-3-3 in CF_3I tested in the dry condition without copper at 150 °C.

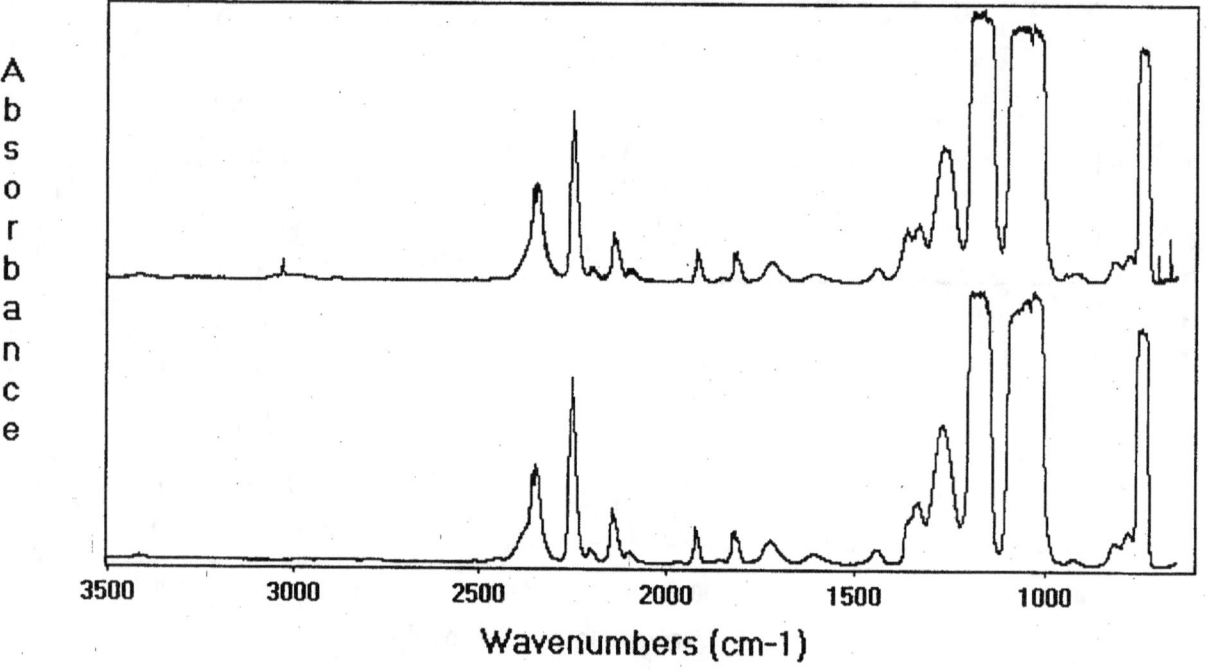

Figure C-14. Initial (lower) and 40 week (upper) spectra for C4130 in CF_3I tested in the dry condition without copper at 150 °C.

7. AGENT STABILITY UNDER STORAGE

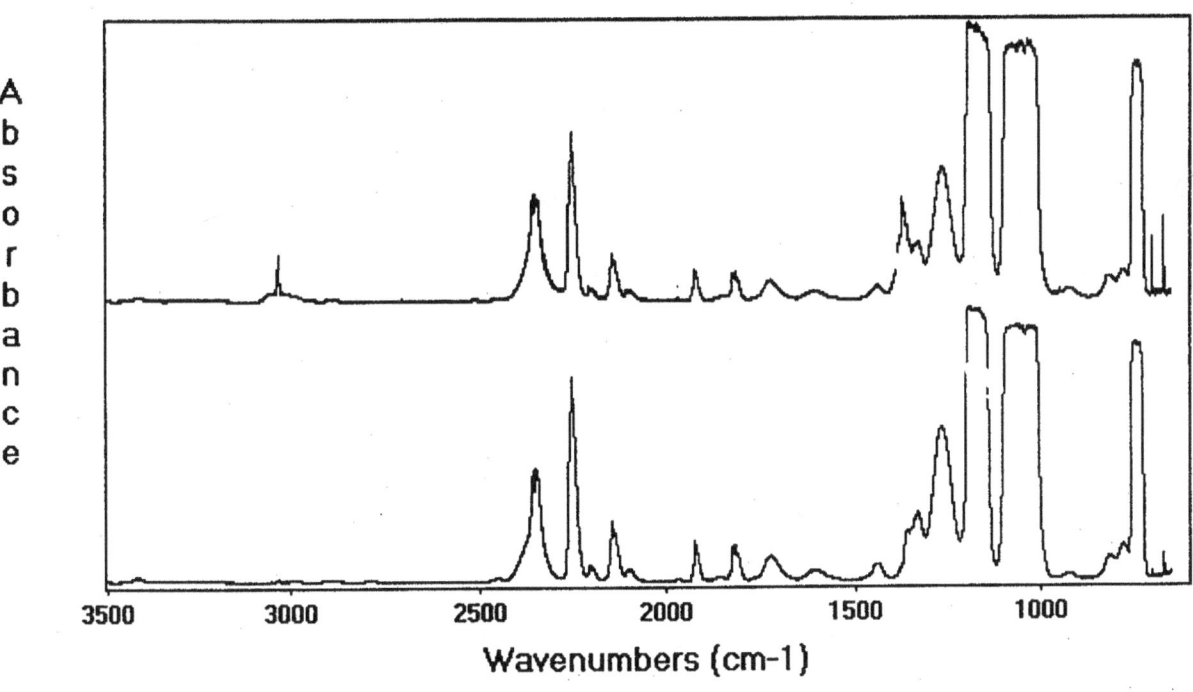

Figure C-15. Initial (lower) and 44 week (upper) spectra for I625 in CF_3I tested in the dry condition without copper at 150 °C.

Appendix D. Initial and Final FTIR Spectra of CF_3I Tested in the Dry Condition with Copper

The spectra in Appendix D are those of the CF_3I. The gas cell pressure for all was 5330 Pa. The lower spectrum is for the initial analysis; the upper spectrum is for the final aged analysis.

7. AGENT STABILITY UNDER STORAGE

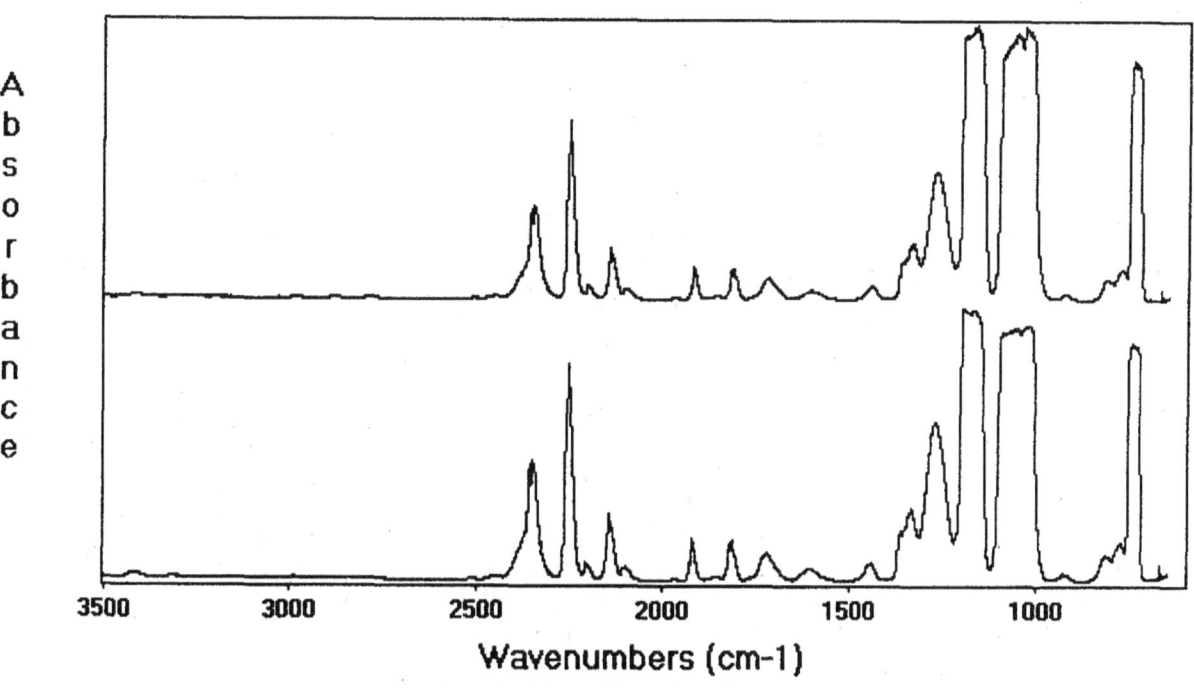

Figure D-1. Initial (lower) and 48 week (upper) spectra for the blank in CF$_3$I tested in the dry condition with copper at 23 °C.

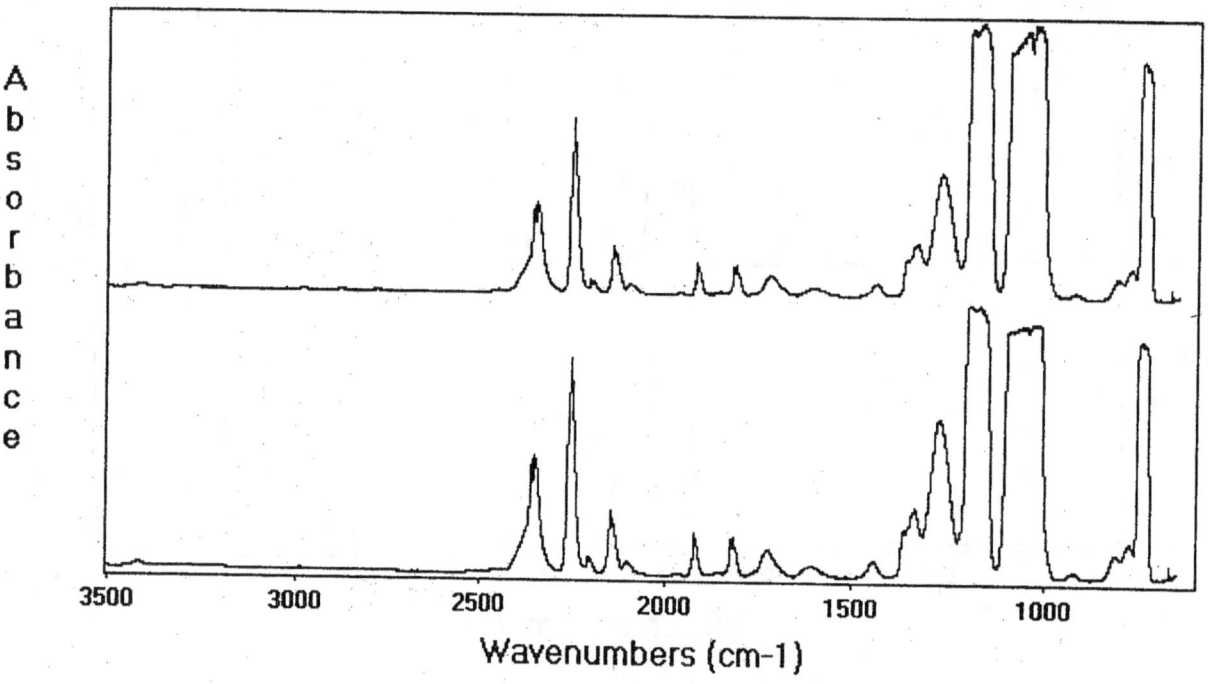

Figure D-2.   Initial (lower) and 48 week (upper) spectra for nitronic 40 in CF$_3$I tested in the dry condition with copper at 23 °C.

7. AGENT STABILITY UNDER STORAGE

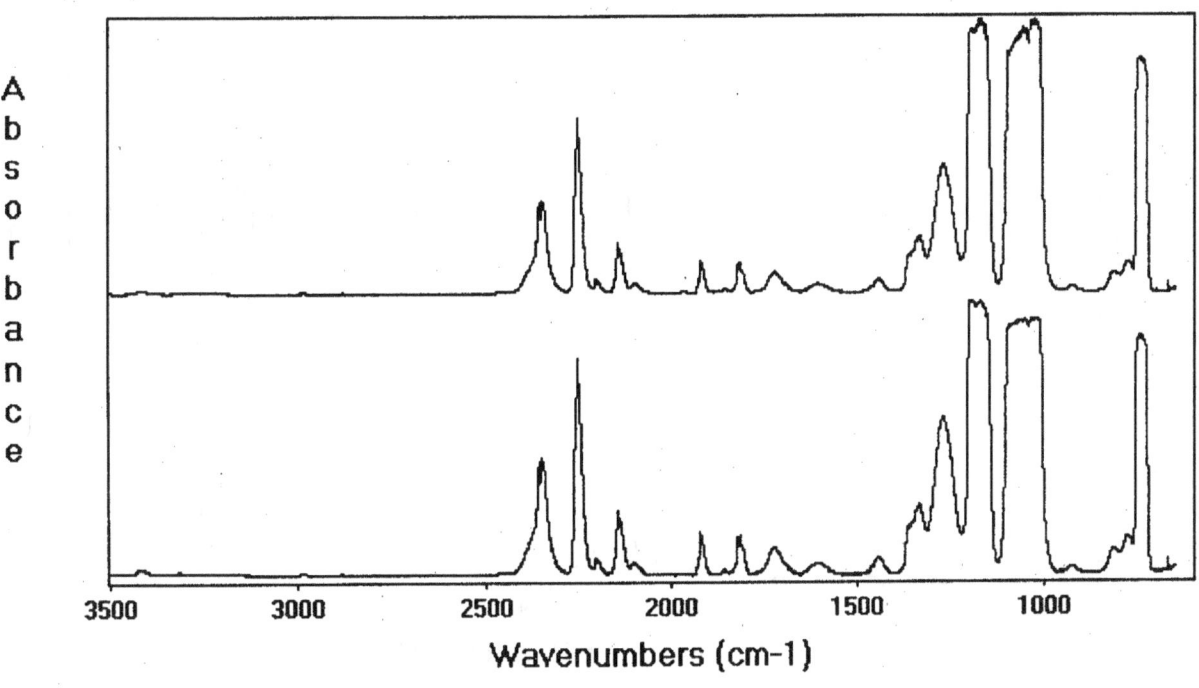

Figure D-3. Initial (lower) and 48 week (upper) spectra for Ti-15-3-3-3 in CF_3I tested in the dry condition with copper at 23 °C.

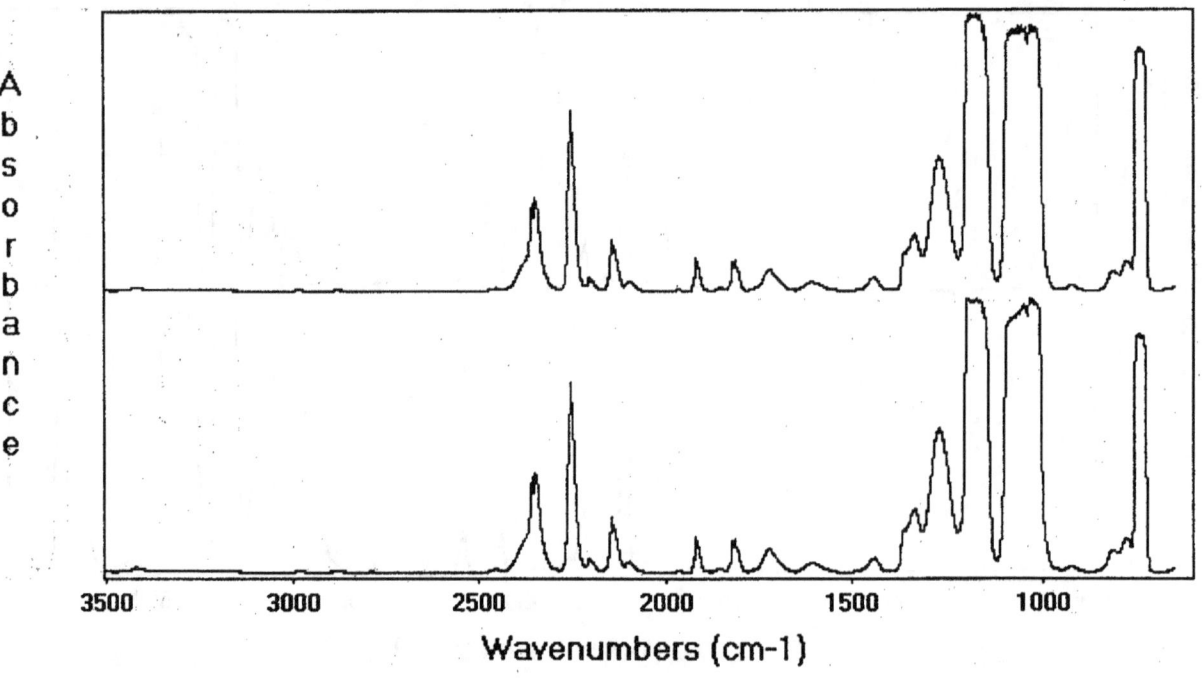

Figure D-4. Initial (lower) and 40 week (upper) spectra for C4130 in CF_3I tested in the dry condition with copper at 23 °C.

7. AGENT STABILITY UNDER STORAGE

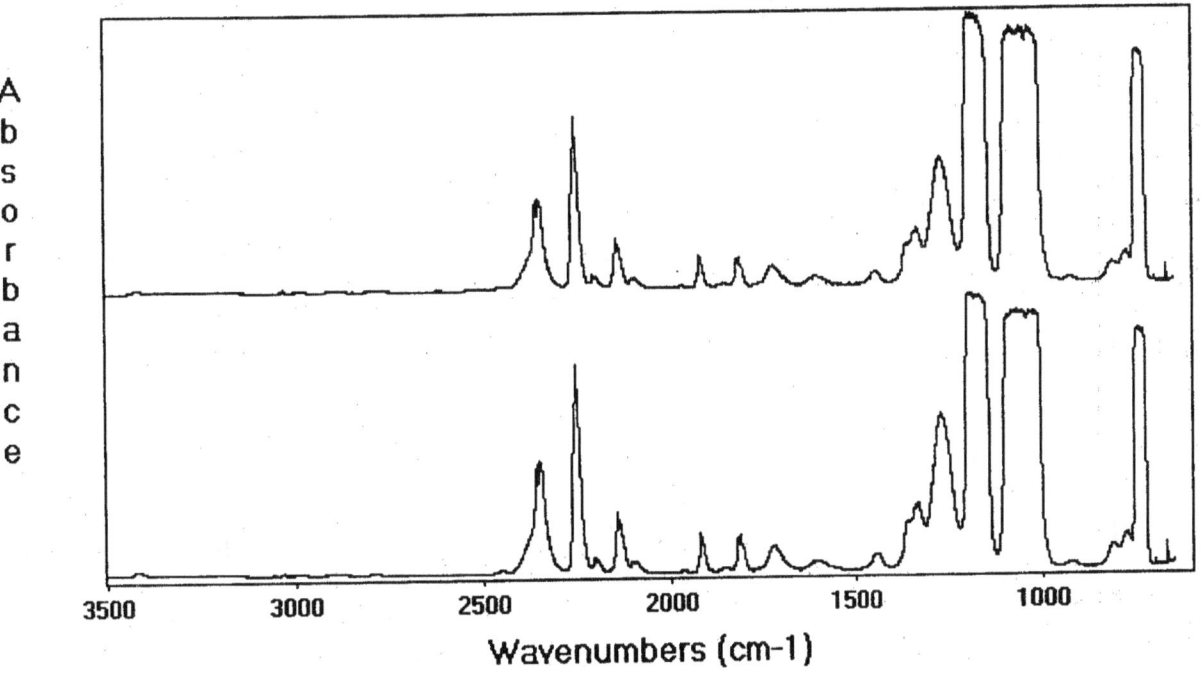

Figure D-5. Initial (lower) and 44 week (upper) spectra for I625 in CF_3I tested in the dry condition with copper at 23 °C.

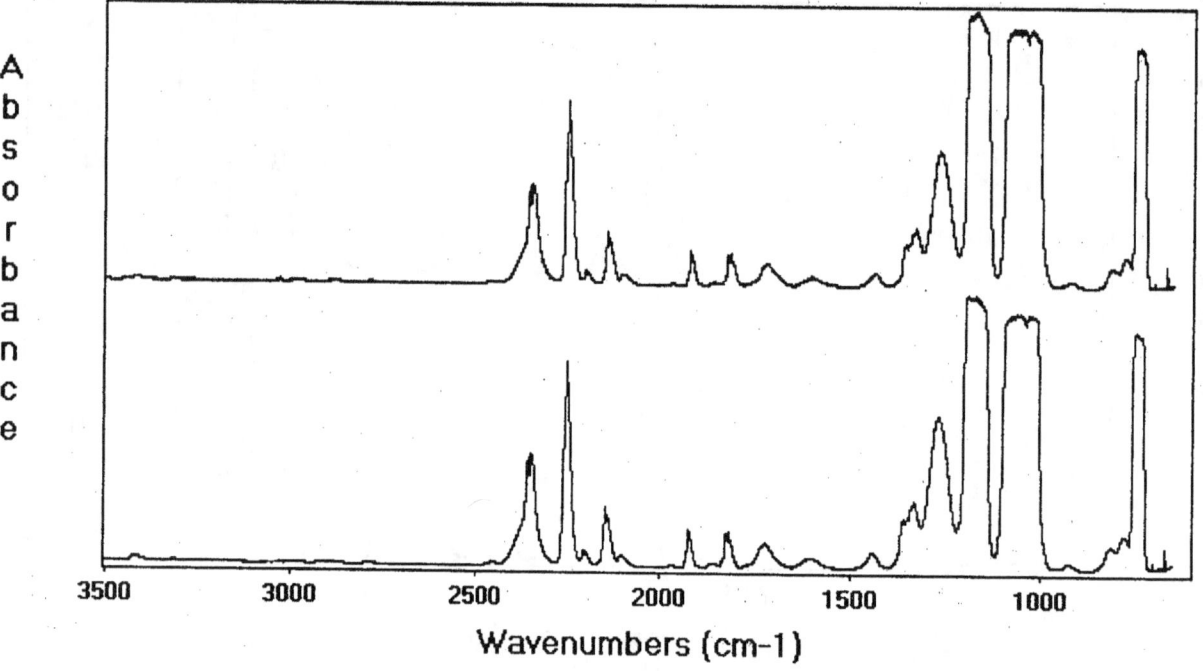

Figure D-6. Initial (lower) and 32 week (upper) spectra for the blank in CF_3I tested in the dry condition with copper at 100 °C.

7. AGENT STABILITY UNDER STORAGE

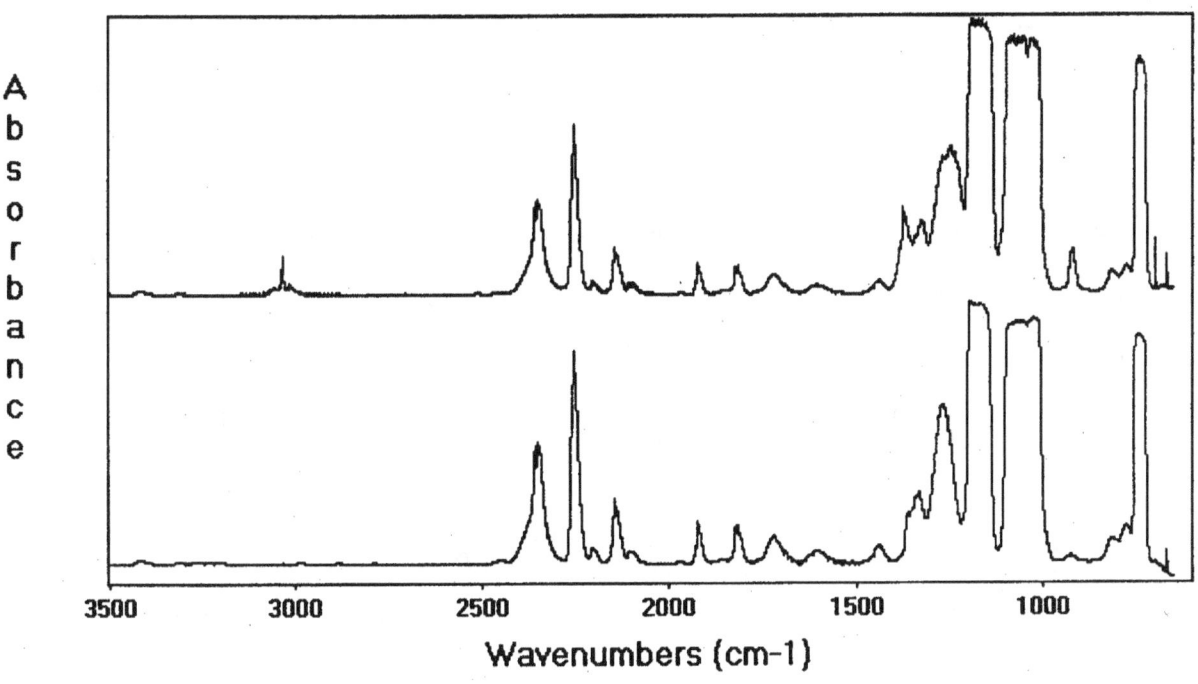

Figure D-7. Initial (lower) and 52 week (upper) spectra for the blank in CF_3I tested in the dry condition with copper at 150 °C.

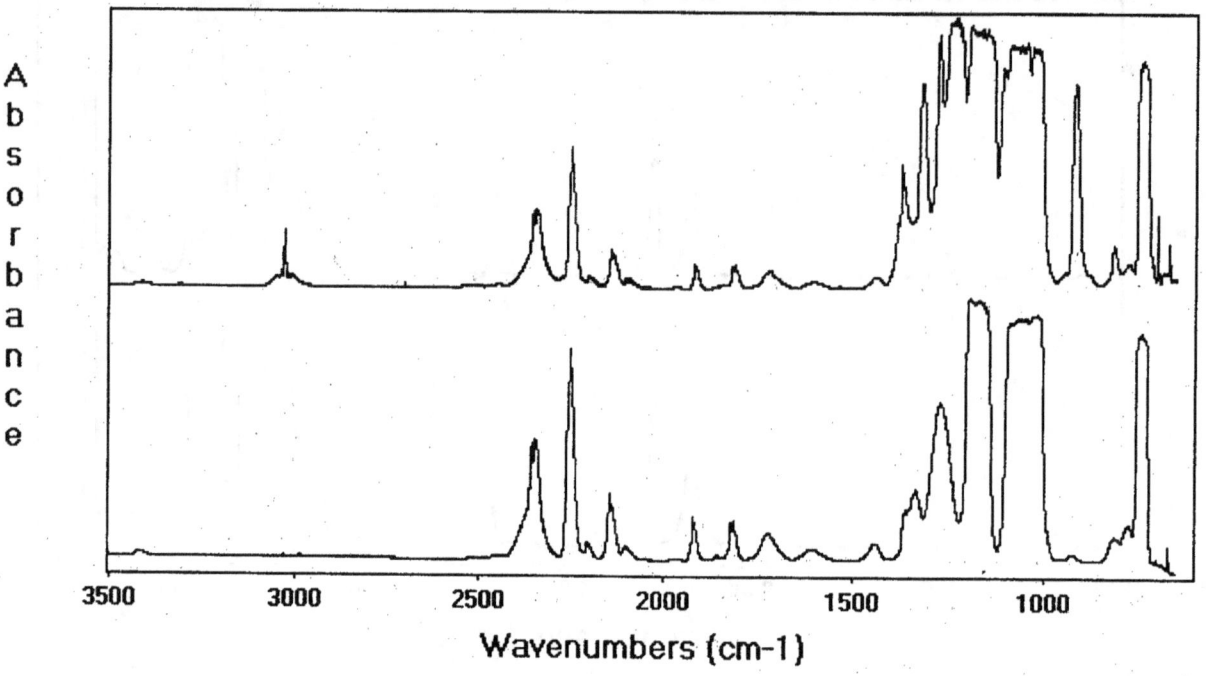

Figure D-8. Initial (lower) and 52 week (upper) spectra for nitronic 40 in CF_3I tested in the dry condition with copper at 150 °C.

7. AGENT STABILITY UNDER STORAGE

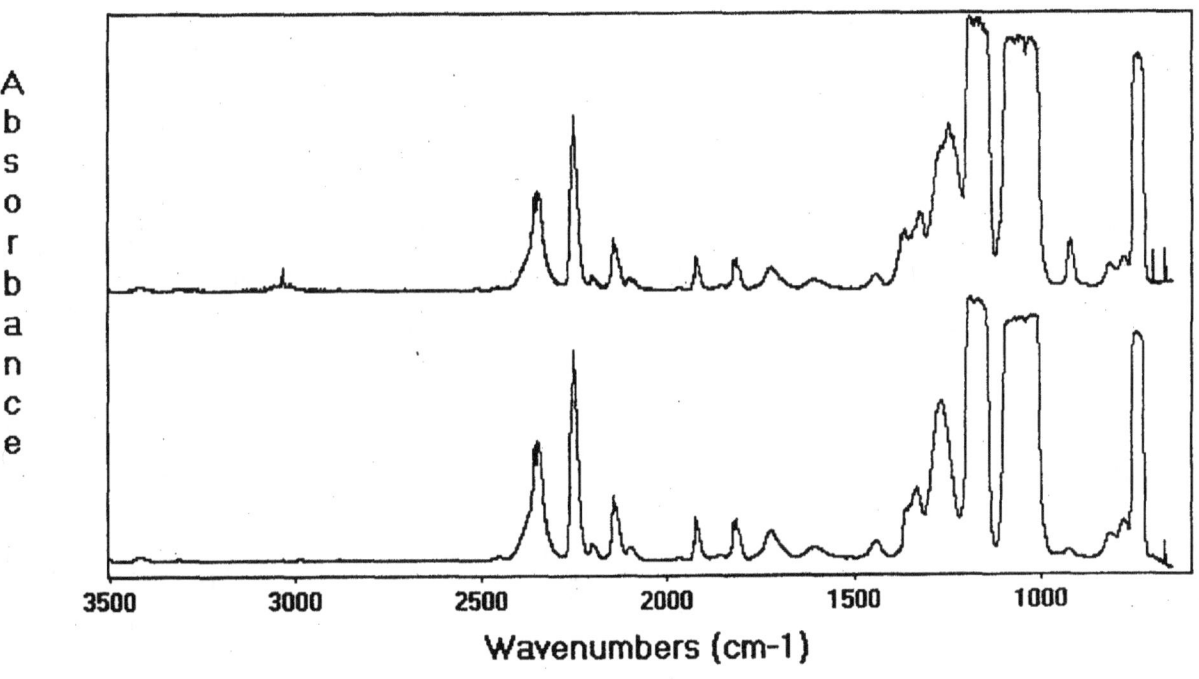

Figure D-9. Initial (lower) and 52 week (upper) spectra for Ti-15-3-3-3 in CF_3I tested in the dry condition with copper at 150 °C.

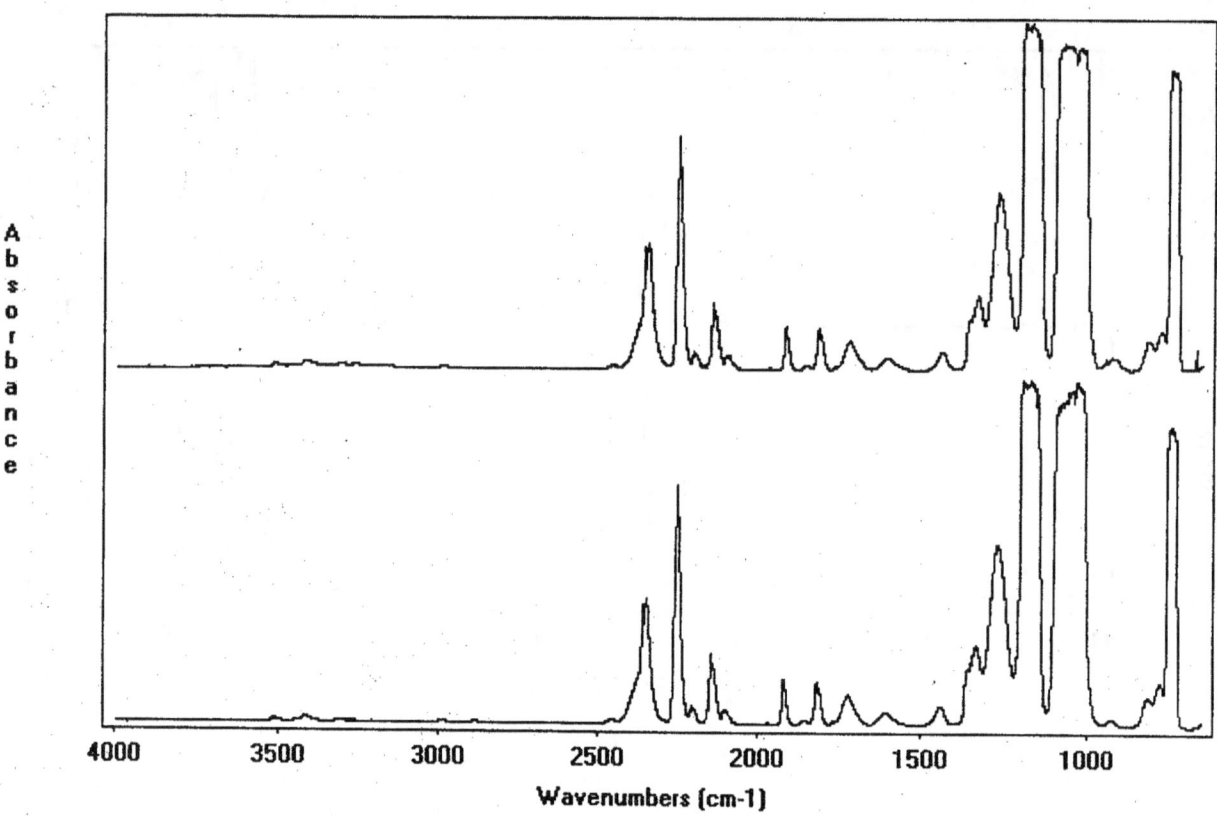

Figure D-10. Initial (lower) and 40 week (upper) spectra for C4130 in CF_3I tested in the dry condition with copper at 150 °C.

7. AGENT STABILITY UNDER STORAGE

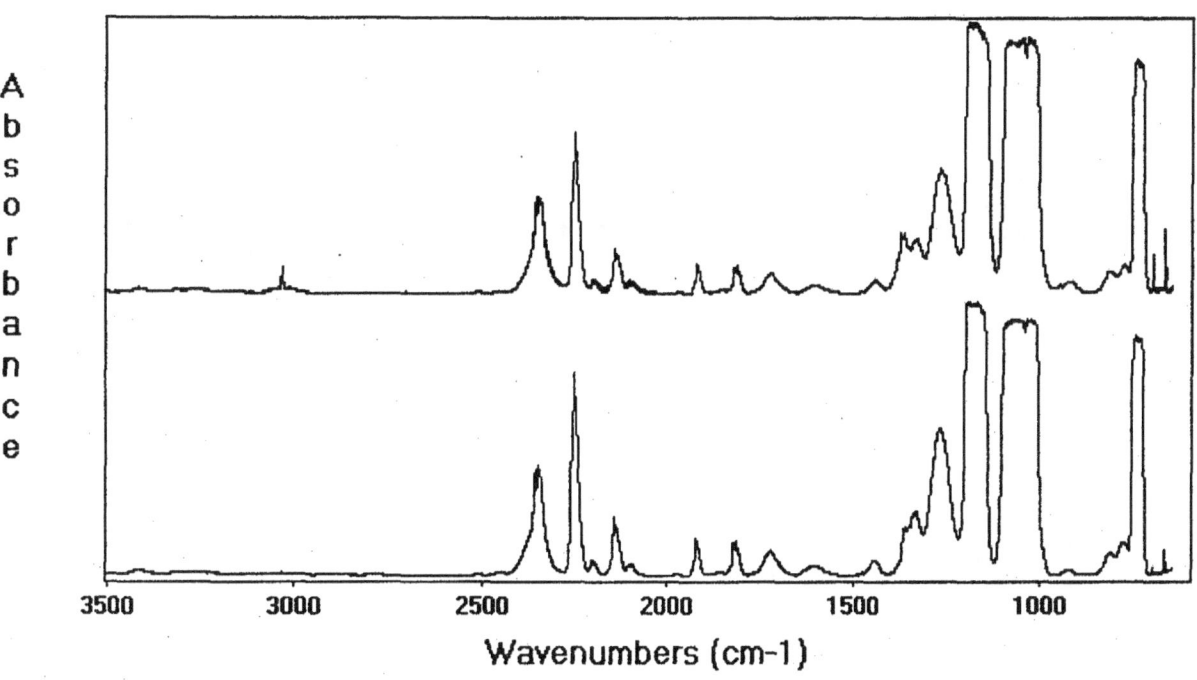

Figure D-11. Initial (lower) and 44 week (upper) spectra for I625 in CF$_3$I tested in the dry condition with copper at 150 °C.

Appendix E. Initial and Final FTIR Spectra of CF$_3$I Tested in the Moist Condition without Copper

The spectra in Appendix E are those of the CF$_3$I. The gas cell pressure for all was 5330 Pa. The lower spectrum is for the initial analysis; the upper spectrum is for the final aged analysis.

7. AGENT STABILITY UNDER STORAGE

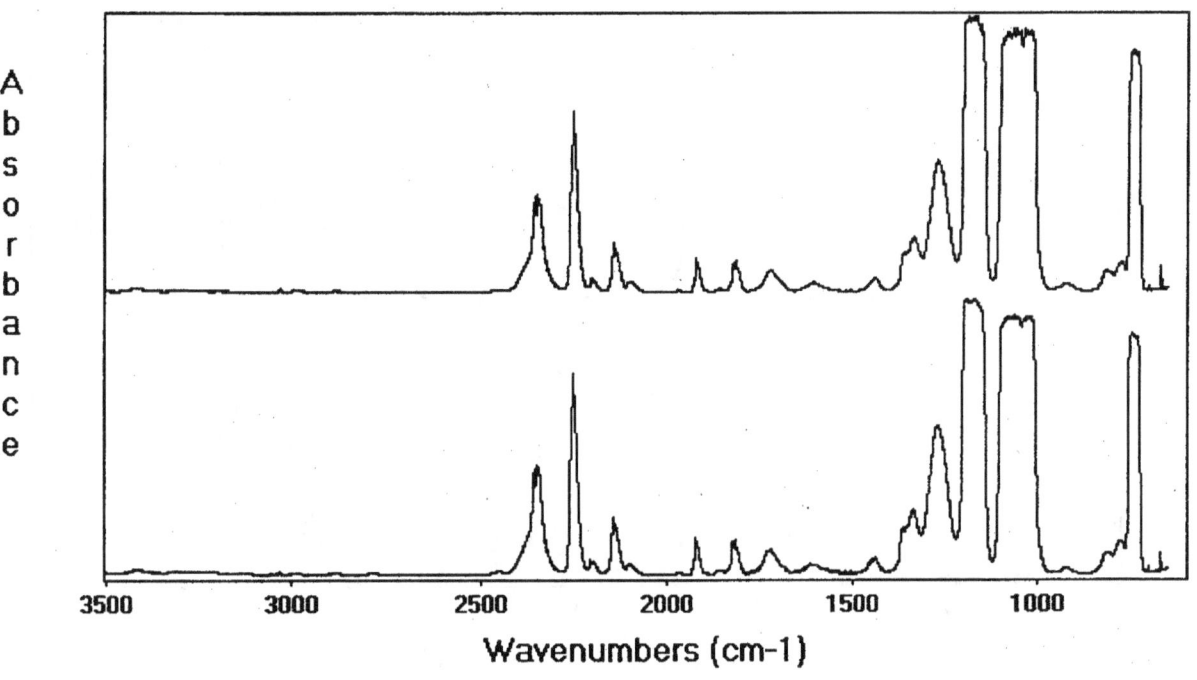

Figure E-1. Initial (lower) and 44 week (upper) spectra for the blank in CF_3I tested in the moist condition without copper at 100 °C.

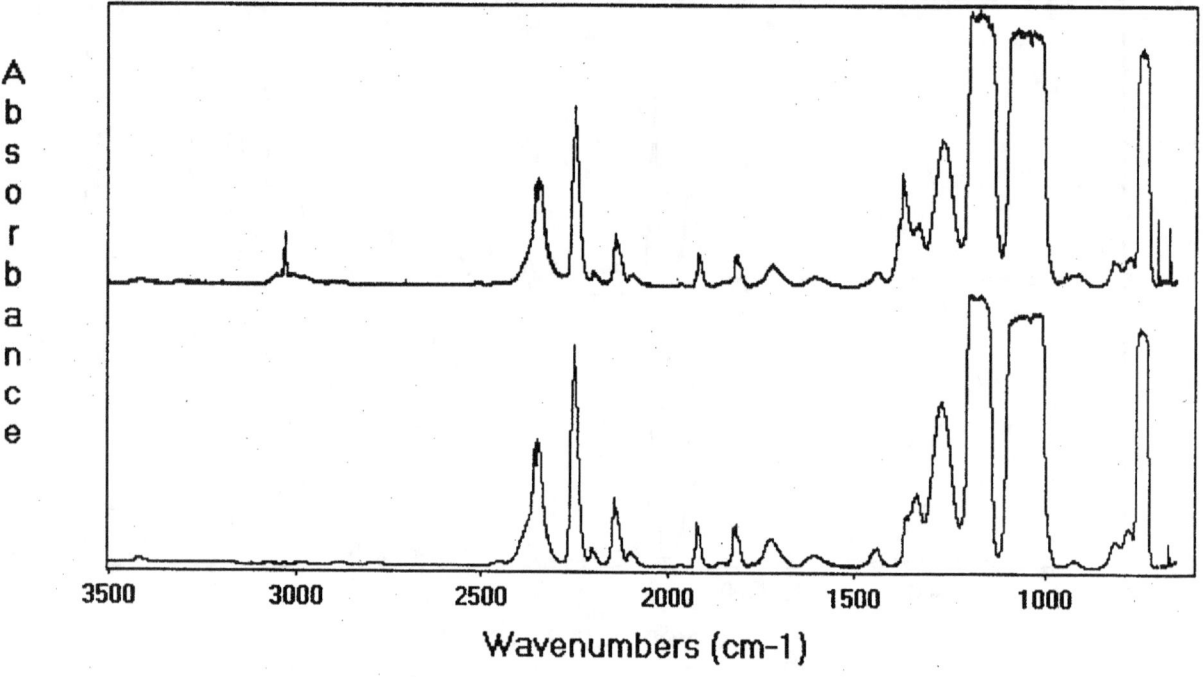

Figure E-2. Initial (lower) and 52 week (upper) spectra for the blank in CF$_3$I tested in the moist condition without copper at 150 °C.

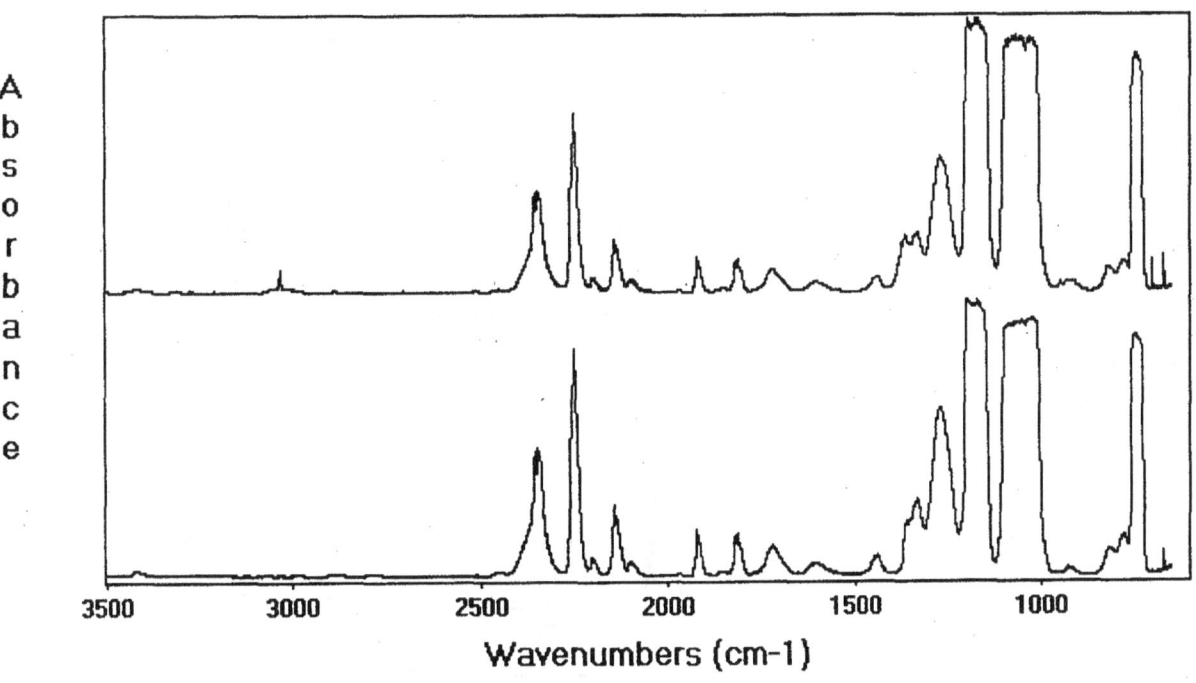

Figure E-3. Initial (lower) and 52 week (upper) spectra for nitronic 40 in CF_3I tested in the moist condition without copper at 150 °C.

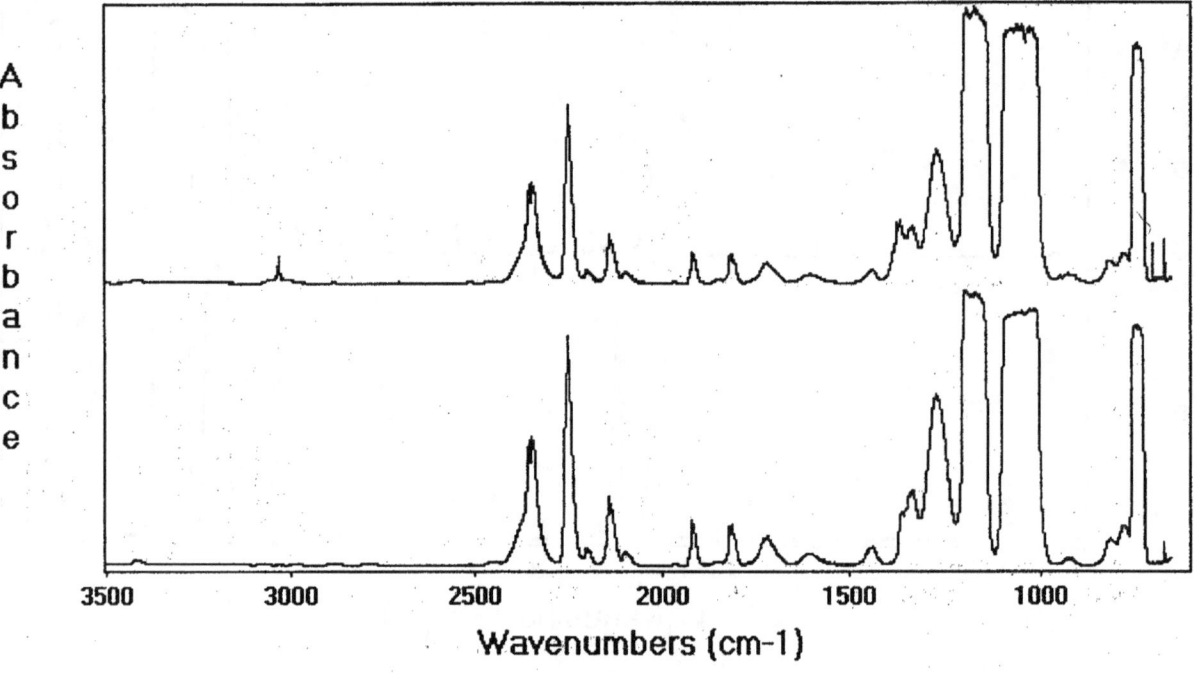

Figure E-4. Initial (lower) and 52 week (upper) spectra for Ti-15-3-3-3 in CF_3I tested in the moist condition without copper at 150 °C.

7. AGENT STABILITY UNDER STORAGE

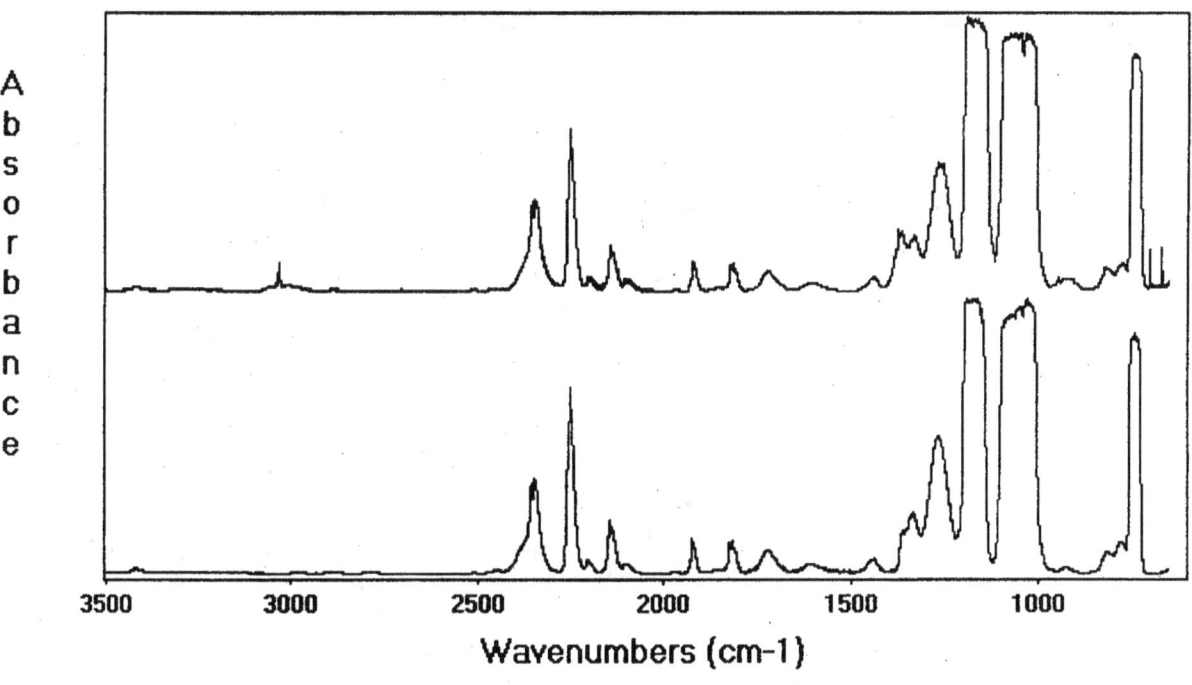

Figure E-5. Initial (lower) and 40 week (upper) spectra for C4130 in CF_3I tested in the moist condition without copper at 150 °C.

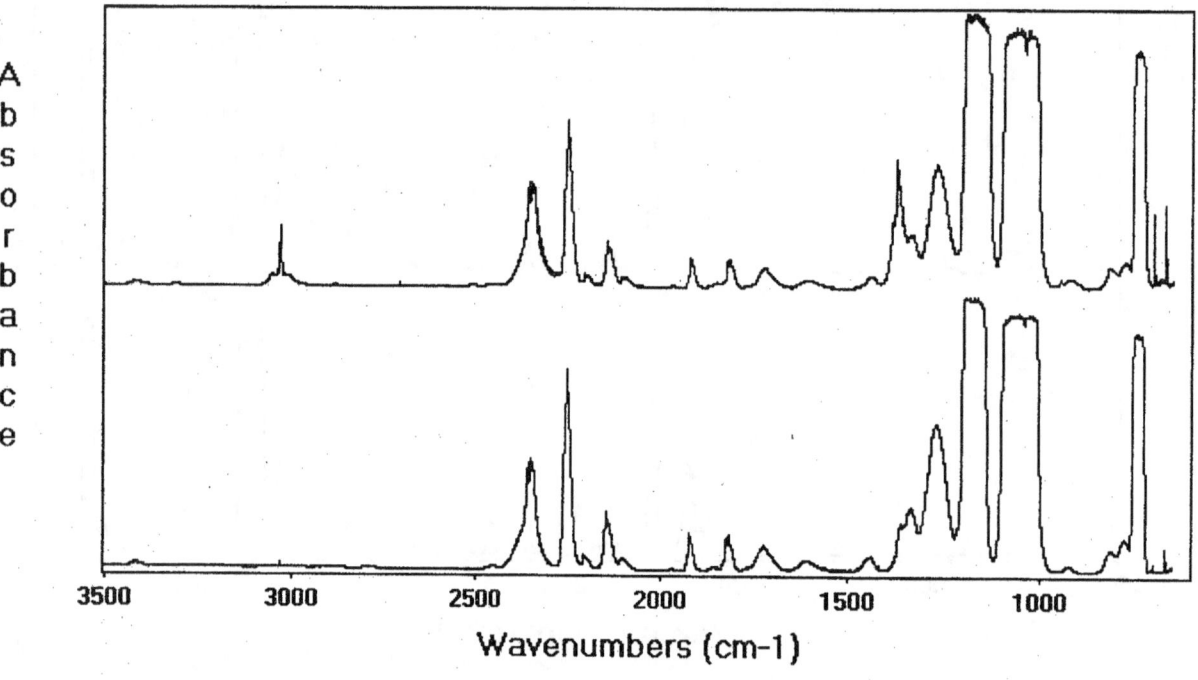

Figure E-6. Initial (lower) and 44 week (upper) spectra for I625 in CF_3I tested in the moist condition without copper at 150 °C.

7. AGENT STABILITY UNDER STORAGE 397

Appendix F. Initial and final FTIR Spectra of CF_3I Tested in the Moist Condition with Copper

The spectra in Appendix F are those of the CF_3I. The gas cell pressure for all was 5330 Pa. The lower spectrum is for the initial analysis; the upper spectrum is for the final aged analysis.

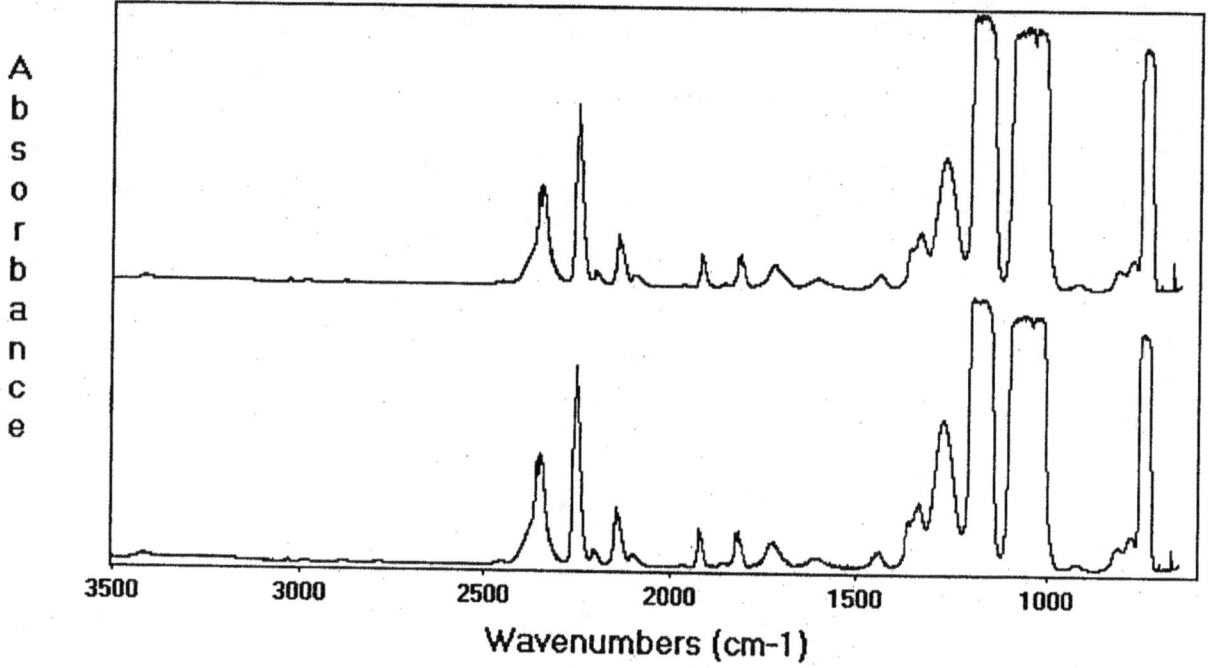

Figure F-1. Initial (lower) and 44 week (upper) spectra for the blank in CF_3I tested in the moist condition with copper at 100 °C.

7. AGENT STABILITY UNDER STORAGE

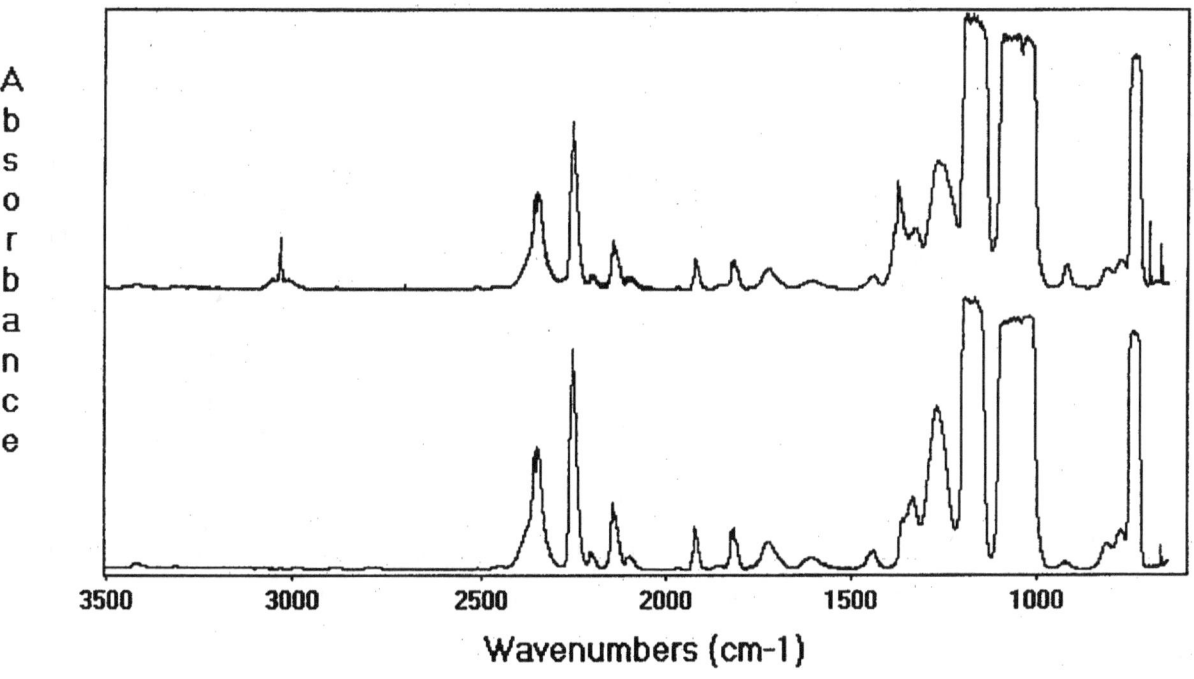

Figure F-2. Initial (lower) and 52 week (upper) spectra for the blank in CF_3I tested in the moist condition with copper at 150 °C.

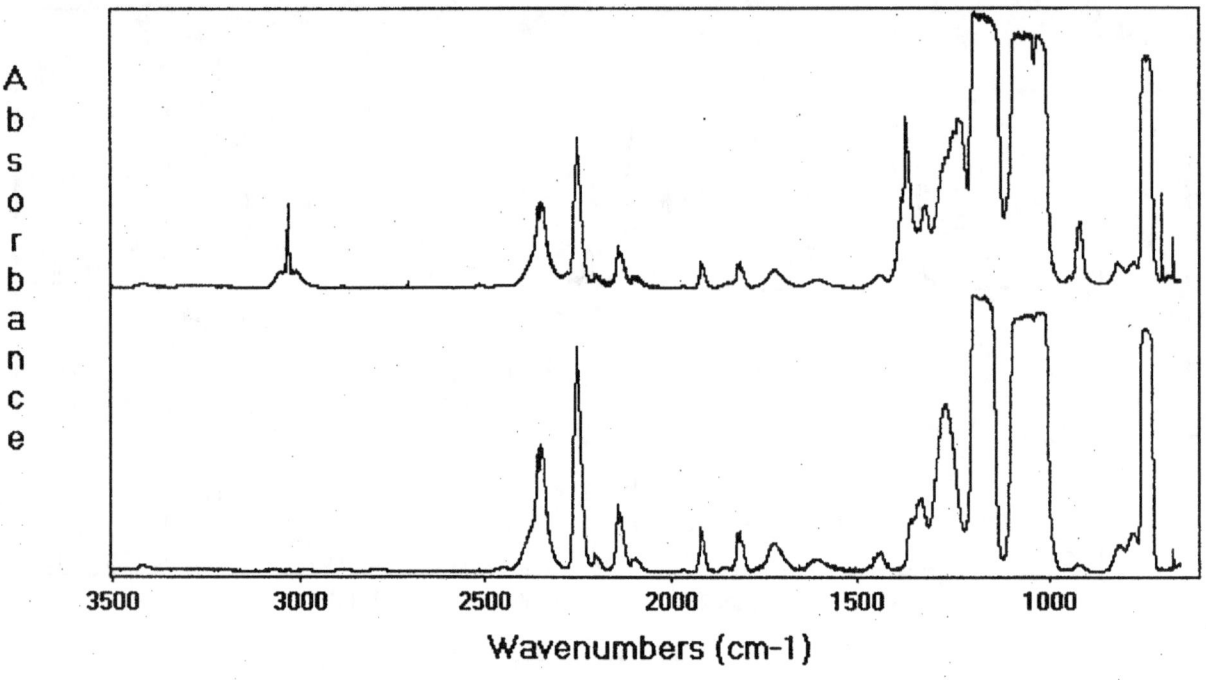

Figure F-3. Initial (lower) and 52 week (upper) spectra for nitronic 40 in CF_3I tested in the moist condition with copper at 150 °C.

7. AGENT STABILITY UNDER STORAGE

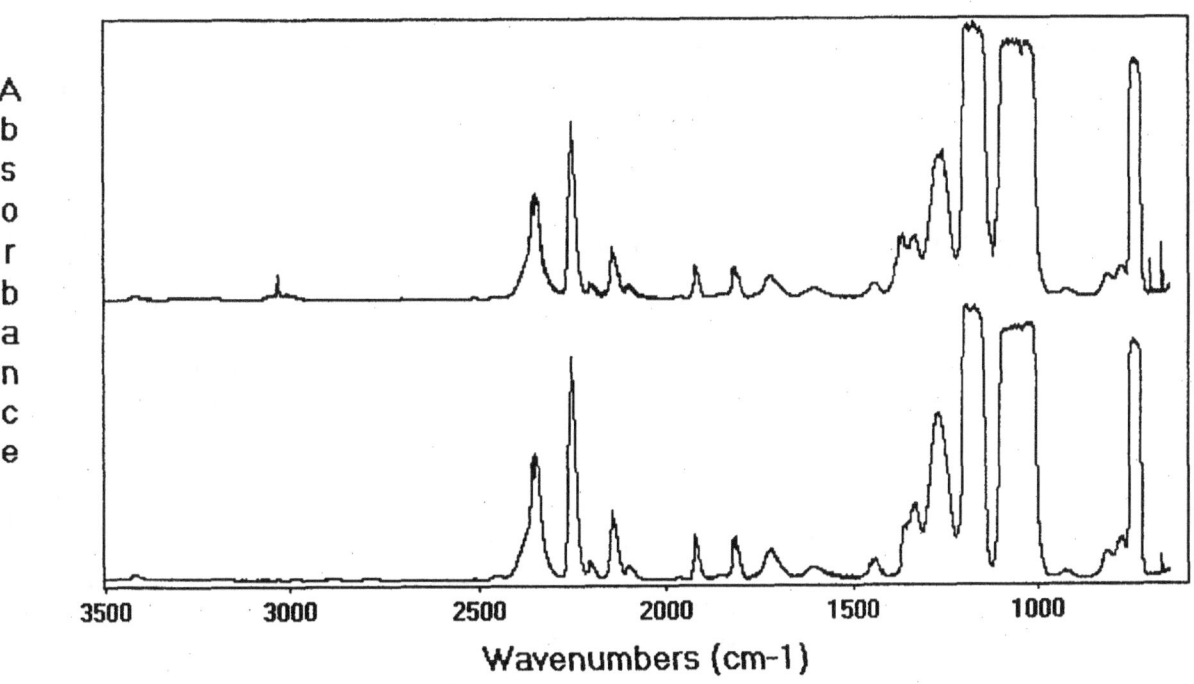

Figure F-4. Initial (lower) and 52 week (upper) spectra for Ti-15-3-3-3 in CF_3I tested in the moist condition with copper at 150 °C.

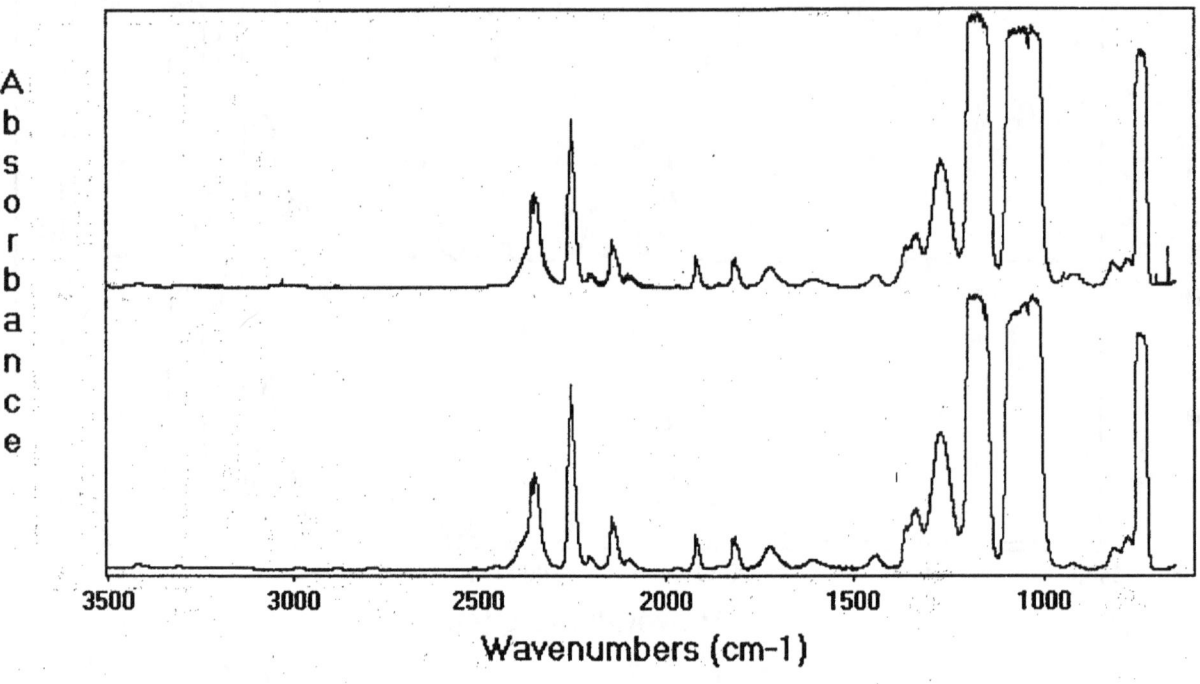

Figure F-5. Initial (lower) and 40 week (upper) spectra for C4130 in CF_3I tested in the moist condition with copper at 150 °C.

7. AGENT STABILITY UNDER STORAGE

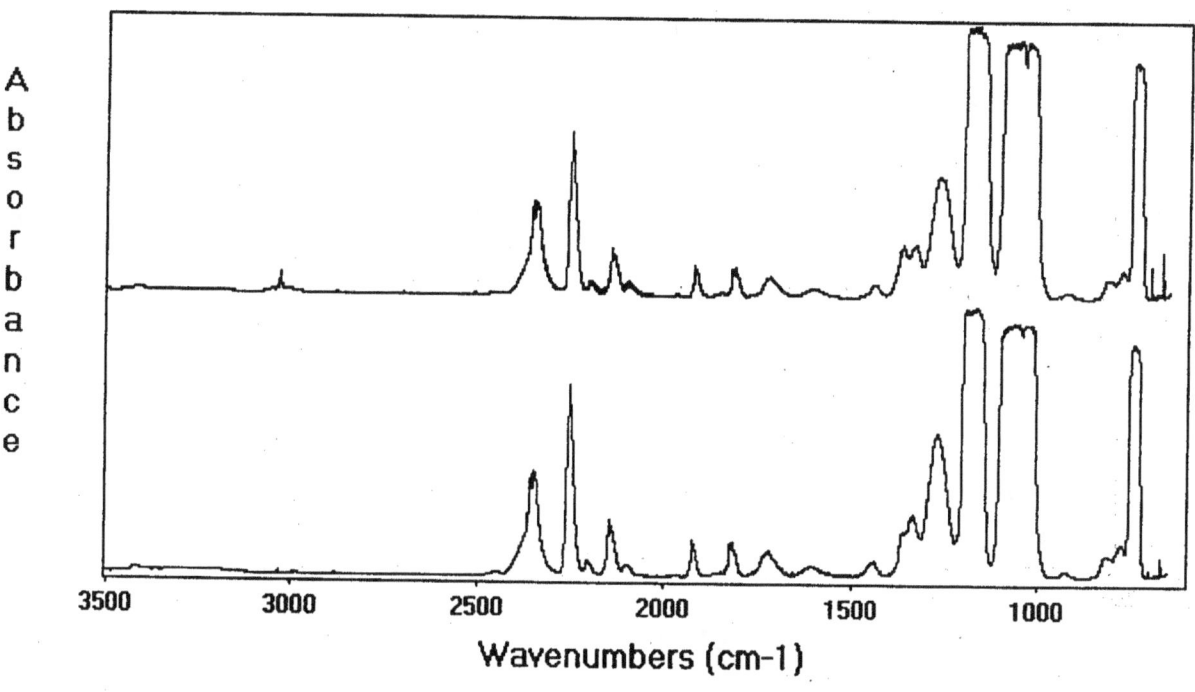

Figure F-6. Initial (lower) and 40 week (upper) spectra for I625 in CF_3I tested in the moist condition with copper at 150 °C.

PART 2

ENGINEERING DESIGN CRITERIA